KU-098-017

transducing the genome

transducing the genome

INFORMATION, ANARCHY, AND REVOLUTION
IN THE BIOMEDICAL SCIENCES

Gary Zweiger

McGraw-Hill

New York • Chicago • San Francisco • Lisbon • London
Madrid • Mexico City • Milan • New Delhi • San Juan
Seoul • Singapore • Sydney • Toronto

Library of Congress Cataloging-in-Publication Data
Zweiger, Gary.
    Transducing the genome : information, anarchy, and revolution in the
biomedical sciences. / Gary Zweiger.
        p. cm.
    Includes index.
    ISBN 0-07-136980-5
    1. Human genome.   2. Human gene mapping.   I. Title.

QH447 .Z94 2000
599.93'5—dc21   00-049628

McGraw-Hill

A Division of The McGraw·Hill Companies

Copyright © 2001 by Gary Zweiger. All rights reserved. Printed in the United
States of America. Except as permitted under the United States Copyright Act
of 1976, no part of this publication may be reproduced or distributed in any form
or by any means, or stored in a data base or retrieval system, without the prior
written permission of the publisher.

1 2 3 4 5 6 7 8 9 0   AGM/AGM   0 6 5 4 3 2 1 0

0-07-136980-5

Printed and bound by Quebecor/Martinsburg.

This publication is designed to provide accurate and authoritative information in
regard to the subject matter covered. It is sold with the understanding that neither the author nor the publisher is engaged in rendering legal, accounting, or
other professional service. If legal advice or other expert assistance is required,
the services of a competent professional person should be sought.

> —From a Declaration of Principles jointly
> adopted by a Committee of the American Bar
> Association and a Committee of Publishers

McGraw-Hill books are available at special quantity discounts to use as premi-
ums and sales promotions, or for use in corporate training programs. For more information, please write to the Director of Special Sales, Professional Publishing,
McGraw-Hill, Two Penn Plaza, New York, NY 10121-2298. Or contact your
local bookstore.

Contents

Acknowledgments

Inspiration for writing this book came from numerous conversations in which scientists and nonscientists alike expressed great interest in the human genome, the Human Genome Project, bioinformatics, gene patents, DNA chips, and other hallmarks of a new age in biology and medicine. There was a sense that history was being made and that biology, medicine, and even our conception of ourselves would be forever changed. For some folks there was tremendous excitement over the possibilities and promises of genomics. For others there was trepidation and concern. However, nearly everyone I spoke with agreed that many of the history-making events and ideas were hidden from the public, and that scientists and general readers alike would benefit from the perspective I had developed over the last decade.

Many of the ideas and insights that I present here came through my work as a biologist in laboratories at Stanford University, Schering-Plough Corporation, Columbia University, and Genentech; in classrooms at various San Francisco Bay area community colleges; consulting for several different biotechnology investment groups; and through scientific, business, and legal work at Incyte Genomics and Agilent Technologies. This work brought me in contact with a vast network of people linked by the common goals of advancing our knowledge of life and improving human health, a perfectly primed source of help for a project such as this.

Several pioneers in the genomics revolution graciously shared their recollections and insights with me. In particular, I thank John Weinstein of the U.S. National Cancer Institute; Leigh Anderson of Large Scale Biology; Leonard Augenlicht of Albert Einstein University; Walter Gilbert of Harvard University; and Randy Scott, Jeff Seilhamer, and Roy Whitfield of Incyte Genomics.

I owe many thanks to coworkers at Incyte Genomics, who were both terrifically forthcoming in sharing their enthusiasm and ideas and tremendously supportive of my extracurricular activities. In particular, I wish to thank Jeanne Loring, Huijun Ring, Roland Somogyi, Stefanie Fuhrman, Tod Klingler, and Tim Johann. I am also especially grateful for the support of my coworkers at Agilent Technologies. In particular, I wish to thank Bill Buffington, Stuart Hwang, Mel Kronick, Doug Amorese, Dick Evans, Paul Wolber, Arlyce Guerry, Linda Lim, Ellen Deering, and Nalini Murdter for their aid and encouragement.

I thank Donny Strosberg of Hybrigenics, Xu Li of Kaiser Permanente, Steve Friend and Chris Roberts of Rosetta InPharmatics, Yiding Wang of Biotech Core, and Brian Ring of Stanford University for images and figures. Thanks to Karen Guerrero of Mendel Biosciences for her encouragement and reviews and for educating me on some of the nuances of patent law.

This book might not have been completed without the steadfast encouragement and expert guidance of Amy Murphy, McGraw-Hill's pioneering "trade" science editor. I also thank Robert Cook-Deegan for his insightful comments.

Finally, I offer my heartfelt thanks to Myriam Zech and Karolyn Zeng for their friendship and advice, Martin and Martha Zweig for their support, my daughter Marissa for her love and inspiration, her mother Andrea Ramirez, and my mother Sheila Peckar.

Introduction

Institutions at the forefront of scientific research host a continual stream of distinguished guests. Scientists from throughout the world come to research centers, such as California's Stanford University, to share ideas and discoveries with their intellectual peers. It is part of an ongoing exchange that embodies the ideals of openness and cooperation characteristic of science in general, and biomedical research in particular.

As a graduate student in the Genetics Department of Stanford's School of Medicine in the late 1980s, I attended several lectures per week by visiting scientists. We heard firsthand accounts of the latest triumphs of molecular biology, of newly discovered molecules, and of their roles in human disease. We heard of elegant, clever, and even heroic efforts to tease apart the molecular architecture of cells and dissect the pathways by which molecular actions lead to healthy physiology or sometimes to disease. And in our own research we did the same. If molecular biology ever had a heyday it was then. The intricate machinery of life was being disassembled one molecule at a time; we marveled at each newly discovered molecule like archaeologists pulling treasures from King Tutankhamen's long-buried tomb. What's more, a few of the molecular treasures were being formulated into life-prolonging medicines by the promising new biotechnology industry.

Nevertheless, there was one guest lecture I attended during this heady time that left me cold. I had been told that Maynard Olson, a highly regarded geneticist then at Washington University in Saint Louis, had helped to develop a powerful new method for identifying genes associated with diseases and traits. But, instead of speaking about this technology or the genes he had discovered, Olson used our attention-filled hour to drone on about a scheme to determine the nucleotide sequence of enormous segments of DNA (deoxyribonucleic acid). The speech was a bore because it had to do with various laboratory devices, automation, and costs. He described technicians (or even graduate students) working on what amounted to an assembly line. He analyzed the costs per technician, as well as costs per base pair of DNA. It was as bad as the rumors we had heard of factory-like sequencing operations in Japan. It all seemed so inelegant, even mindless.

It was not as if DNA wasn't inherently interesting. DNA was (and still is) at the center of biology's galaxy of molecules. Its sequence dictates the composition, and thus the function, of hundreds of thousands of other molecules. However, in the past DNA sequencing had almost always been directed at pieces of DNA that had been implicated in particular biological functions, relevant to specific scientific queries. What seemed so distasteful about a DNA sequencing factory was that it would presumably spew out huge amounts of DNA sequence data indiscriminately. Its product would not be the long-sought answer to a pressing scientific puzzle, but merely enormous strings of letters, As, Cs, Ts, and Gs (the abbreviations for the four nucleotides that make up DNA). Only a computer could manage the tremendous amount of data that a DNA sequencing factory would produce. And computers were not then of great interest to us.

In the late 1980s most biologists had little use for computers other than to compare DNA sequences and communicate with each other over a network that later evolved into the Internet. Only a few of them embraced the new technology the way that scientists in other disciplines had. Biologists were compelled by an interest in organic systems, not electronic systems, and most relished the hands-on experience of the laboratory or the field. Most biologists considered computers as being just another piece of laboratory equipment, although some perceived them as a threat to their culture. One day a graduate student I knew who had decided to embark on research that was entirely computer-based found himself in an elevator with the venerable Arthur Kornberg, a biochemist who had won a Nobel prize for identifying and characterizing the molecules that replicate DNA. Probably more than anyone else, Kornberg was responsible for establishing Stanford's world-renowned Biochemistry Department and for creating the U.S. government's peer-review system for distributing research grants. He would later author a book entitled *For the Love of Enzymes*. He was also a curmudgeon, and when the elevator doors closed upon him and the unfortunate graduate student he reportedly went into a finger-wagging tirade about how computation would never be able to replace the experiments that this group did in the laboratory.

Which brings us to the subject of this book. In the late 1980s we were at the dawn of a major transformation within the biomedical sciences. I didn't realize it at the time, but Olson's lecture and my colleague's commitment to computation were portents of exciting and significant things to come. The life sciences are now undergoing a dramatic shift from single-gene studies to experiments involving thousands of genes at a time, from small-scale academic studies to industrial-scale ones, and from a molecular approach to life to one that is information-based and computer-intensive. This transformation has already had a profound effect on life sciences research. It is beginning to have a profound effect on medicine and agriculture. In addition, it is likely to bring about significant changes in our understanding of ourselves, of other human beings, and of other living creatures. Change can be a raging bull, frightening in its power and unpredictability. The pages that follow are an attempt to grasp the bull by its horns, to understand the nature and origin of the "New Biology," and to deliver this beast to you, the readers.

Biology is being reborn as an information science, a progeny of the Information Age. As information scientists, biologists concern themselves with the messages that sustain life, such as the intricate series of signals that tell a fertilized egg to develop into a full-grown organism, or the orchestrated response the immune system makes to an invading pathogen. Molecules convey information, and it is their messages that are of paramount importance. Each molecule interacts with a set of other molecules and each set communicates with another set, such that all are interconnected. Networks of molecules give rise to cells; networks of cells produce multicellular organisms; networks of people bring about cultures and societies; and networks of species encompass ecosystems. Life is a web and the web is life.

Ironically, it was the euphoria for molecules that touched off this scientific revolution. In the 1980s only a tiny percentage of the millions of different molecular components of living beings was known. In order to gain access to these molecules, a new science and even a new industry had to be created. *Genomics* is the development and application of research tools that uncover and analyze thousands of different molecules at a time. This new approach to biology has been so successful that universities have created entire departments devoted to it, and all major pharmaceutical companies now have large genomics divisions. Genomics has granted biologists unprecedented access to the molecules of life, but what will be described here is more than just a technological revolution. Through genomics massive amounts of biological information can be converted into an electronic format. This directly links the life sciences to the information sciences, thereby facilitating a dramatically new framework for understanding life.

Information is a message, a bit of news. It may be encoded or decoded. It may be conveyed by smoke signals, pictures, sound waves, electromagnetic waves, or innumerous other media, but the information itself is not made of anything. It has no mass. Furthermore, information always has a sender and an intended receiver. This implies an underlying intent, meaning, or purpose. Information theory thus may seem unfit for the cold objectivism of science. The focus of the information sciences, however, is not so much on message content, but rather on how messages are conveyed, processed, and stored.

Advances in this area have been great and have helped to propel the remarkable development of the computer and telecommunication industries. Could these forces be harnessed to better understand the human body and to improve human health? The gene, as the Czech monk Gregor Mendel originally conceived it, is a heritable unit of information passed from parent to offspring. Mathematical laws describing the transmission of genes were described a century ago, long before the physical nature of genes was finally determined in the 1950s. At the core of the molecular network of every living organism is a *genome*, a repository of heritable information that is typically distributed throughout all the cells of the organism. "They are law code and executive power—or to use another simile, they are architect's plan and builder's code in one," explained the renowned physicist Erwin Schrödinger in his famous 1943 lecture entitled "What Is Life?" The genome consists of the complete set of genes of an organism. In humans it is encoded by a sequence of over three billion nucleotides (the molecular subunits of DNA). This information plays such a central role that it has been called the "Book of Life," the "Code of Codes," and biology's "Holy Grail," "Rosetta Stone," and "Periodic Chart." We will see how this information became fodder for modern information technology and the economics of the Information Age.

The Human Genome Project, a government- and foundation-sponsored plan to map and sequence the human genome, and several privately funded sequencing initiatives have been hugely successful. The identity and position of nearly all 3.1 billion nucleotides have now been revealed. This knowledge of the source code of *Homo sapiens*, a glorious achievement and a landmark in the history of humankind, did not come without tension and controversy. Genomics pioneer Craig Venter received harsh criticism, first when he left a government laboratory to pursue his plan for rapid gene discovery with private funds, and later when he founded a company, Celera Genomics, whose primary mission was to sequence the human genome before all others. Noncommercial and commercial interests, represented mainly by Celera and Incyte Genomics, have clashed and competed in a vigorous race to identify human genes. Efforts to claim these genes as intellectual property have been met with fierce criticism.

Interestingly, both the commercial and noncommercial initiatives have also thoroughly relied upon each other. The Human Genome Project would be inconceivable without the automated sequencing machines developed by Michael Hunkapiller and colleagues at Applied Biosystems Inc. On the other hand, Hunkapiller's work originated in Leroy Hood's government-backed laboratory at the California Institute of Technology. In the chapters that follow I examine the forces, people, and ideas that have been propelling the search for human genes, shaping the genomics industry, and creating a dramatically new understanding of life.

I also examine the human genome itself, the forces that have shaped it, and what it may reveal about ourselves. The Human Genome Project and other sequencing initiatives provide us with the information content of the genome, a starting point for countless new analyses. Within the three-billion-letter sequence we can detect the remnants of genes that helped our distant ancestors survive and the sequences that set us apart from other species. Using "DNA chips" we can detect the thousands of minute variations that make each of us genetically unique, and with the help of sophisticated computer algorithms we can now determine which sets of variations lead to disease or to adverse reactions to particular medical treatments. Other algorithms help us understand how a complex network of molecular messages coordinates the growth of tissue and how perturbations in the network may lead to diseases, such as cancer.

To aid in storing and analyzing genomic data, Celera Genomics has a bank of computers capable of manipulating 50 terabytes of data, enough to hold the contents of the Library of Congress five times over, while Incyte's Linux-run computer farm manages a mind-boggling 75 terabytes of data. By transducing the genome—transferring its information content into an electronic format—we open up tremendous new opportunities to know ourselves and better our lives. In this communication about communications, I will consider the molecular language of life, the messages that flow within us, including those that signal disease. I will explain how they are intercepted, transduced into electrical signals, and analyzed, and will describe our efforts to use these analyses responsibly, to respond to disease conditions with carefully constructed molecular replies (i.e., medicine).

Humankind, our collective identity, is like a child forever growing up. We seem to progressively acquire more power and greater responsibility. Our actions now have a profound effect on the environment and on virtually all forms of life. We have become the stewards of planet Earth. By transducing the genome we acquire even greater responsibilities, becoming stewards of our own genome (philosophically, a rather perplexing notion). In the chapters that follow I will describe who is generating and applying knowledge of our genome—and why. I hope that this will help us to better evaluate our collective interests and determine how these interests can be best supported.

1

Cancer, Computers, and a "List-Based" Biology

Science is about providing truthful explanations and trustworthy predictions to an otherwise poorly understood and unpredictable world. Among the greatest of scientific challenges is cancer. We've been in a state of declared war with cancer for decades, yet despite rising expenditures on research (close to $6 billion in 2000 in the United States alone) and treatment (about $40 billion in 2000 in the U.S.), cancer remains a mysterious and seemingly indiscriminant killer. Each year about 10 million people learn that they have cancer (1.2 million in the U.S.) and 7.2 million succumb to it (600,000 in the U.S.), often after much suffering and pain.

Cancer is a group of diseases characterized by uncontrolled and insidious cell growth. The diseases' unpredictable course and uncertain response to treatment are particularly vexing. Cancer patients show tremendous variation in their response to treatment, from miraculous recovery to sudden death. This uncertainty is heart-wrenching for patients, their loved ones, and their caregivers. Moreover, there is little certainty about what will trigger the onset of uncontrolled cell growth. With cancer, far too frequently, one feels that one's fate relies on nothing more than a roll of the dice. If your aunt and your grandmother had bladder cancer, then you may have a 2.6-fold greater chance of getting it than otherwise. If you butter your bread you may be twice as likely to succumb to a sarcoma than you will be if you use jam. A particular chemotherapeutic drug may give you a 40 percent chance of surviving breast cancer—or only a 10 percent chance if you already failed therapy with another chemotherapeutic drug. Clearly, cancers are complex diseases with multiple factors (both internal and external) affecting disease onset and progression. And clearly, despite tremendous advances, science has yet to win any battle that can be seen as decisive in the war against cancer.

Perhaps, a revolutionary new approach, a new framework of thinking about biology and medicine, will allow us to demystify cancer and bring about a decisive victory. The outlines of what may prove to be a successful new scientific paradigm are already being drawn.

Knowing one's enemy often helps in defeating one's enemy, and in the early 1980s Leigh Anderson, John Taylor, and colleagues at Argonne National Laboratory in Illinois pioneered a new method for knowing human cancers. Indeed, it was a new way of knowing all types of cells. Previous classification schemes relied on visual inspection of cells under a microscope or on the detection of particular molecules (known as markers) on the surface of the cells. Such techniques could be used to place cancers into broad categories. A kidney tumor could be distinguished from one derived from the nearby adrenal gland, for example. However, a specific tumor that might respond well to a particular chemotherapeutic agent could often not be distinguished from one that would respond poorly. A tumor that was likely to spread to other parts of the body (metastasize) often could not be distinguished from one that was not. They often looked the same under a microscope and had the same markers. The Argonne team took a deeper look. They broke open tumor cells and surveyed their molecular components. More precisely, they surveyed their full complement of proteins. Proteins are the workhorses of the cell. They provide cell structure, catalyze chemical reactions, and are more directly responsible for cell function than any other class of molecules. Inherent differences in tumors' responses to treatment would, presumably, be reflected by differences in their respective protein compositions.[1]

Anderson, who holds degrees in both physics and molecular biology, was skilled in a technique known as *two-dimensional gel electrophoresis*. In this procedure the full set of proteins from a group of cells is spread out on a rectangular gel through the application of an electrical current in one direction and a chemical gradient in the orthogonal (perpendicular) direction. The proteins are radioactively labeled, and the intensity of the emitted radiation reflects their relative abundance (their so-called "level of expression"). X-ray film converts this radiation into a constellation of spots, where each spot represents a distinct protein, and the size and intensity of each spot corresponds with the relative abundance of the underlying protein (see Fig. 1.1). Each cell type produces a distinct constellation, a signature pattern of spots. If one could correlate particular patterns with particular cell actions,

FIGURE 1.1   Protein constellation produced by two-dimensional gel electrophoresis. Image from LifeExpress Protein™, a collaborative proteomics database produced by Incyte Genomics and Oxford GlycoSciences.

then one would have a powerful new way of classifying cell types. Anderson and colleagues wrote:

> 2-D protein patterns contain large amounts of quantitative data that directly reflect the functional status of cells. Although human observers are capable of searching such data for simple markers correlated with the available external information, global analysis (i.e., examination of the entire data) for complex patterns of change is extremely difficult. Differentiation, neoplastic transformation [cancer], and some drug effects are known to involve complex changes, and thus there is a requirement to develop an approach capable of dealing with data of this type.[2]

Taylor, a computer scientist by training, had the skills necessary to make this new classification scheme work. He took the protein identities (by relative position on the film) and their intensities and transduced them into an electronic format with an image scanner. This information was then captured in an electronic database. There were 285 proteins that could be identified in each of the five tumor cell samples. Measurements were taken three or four times, increasing the database available for subsequent analysis to 4560 protein expression values. (With numbers like these, one can readily see why an electronic base is essential. Imagine scanning 4560 spots or numbers by eye!) Armed with this data, the Argonne group embarked on research that was entirely computer-based.

If only one protein is assayed, one can readily imagine a classification scheme derived from a simple number line plot. A point representing each cell sample is plotted on the number line at the position that corresponds to the level of the assayed protein. Cell samples are then classified or grouped according to where they lie on the line. This is how classical tumor marker assays work. The marker (which is usually a protein on the surface of the cell) is either present at or above some level or it is not, and the tumor is classified accordingly.

With two proteins, one can plot tumor cell samples as points in two-dimensional space. For each cell sample, the x-coordinate is determined by the level of one protein, and the y-coordinate is determined by the level of the second protein. A cell sample could then be classified as being high in protein 1 and low in protein 2, or high in both proteins 1 and 2, etc. Thus, having two data points per tumor enables more categories than having just one data point. However, variations in just one or two proteins may not be sufficient to distinguish among closely related cell types, particularly if one does not have any prior indication of which proteins are most informative. The Argonne group had the benefit of 285 protein identifiers for each tumor cell sample.

Mathematically, each cell sample could be considered of as a point in 285-dimensional space. Our minds may have trouble imagining so many dimensions, but there are well-established mathematical methods that can readily make use of such information. A computer program instantly sorted Anderson and Taylor's five tumor cells samples into categories based on their 4560 protein values. Another program created a dendrogram or tree diagram that displayed the relationships among the five tumor cell types. A powerful new method of cell classification had been born.

The five cancer cell culture protein patterns were intended to be a small portion of a potential database of thousands of different cell cultures and tissue profiles. Leigh Anderson and his father Norman Anderson, also of the Argonne National Laboratory, had a grand scheme to catalogue and compare virtually all human proteins. Since the late 1970s they had campaigned tirelessly for government and scientific support for the initiative, which they called the Human Protein Index. The Andersons had envisioned a reference database that every practicing physician, pathologist, clinical chemist, and biomedical researcher could access by satellite.[3] Their two-dimensional gel results would be compared to protein constellations in this database, which would include links to relevant research reports. The Andersons had also planned a computer system that would manage this information and aid in its interpretation. They called the would-be system TYCHO, after Tycho Brahe, the famous Danish astronomer who meticulously catalogued the positions of stars and planets in the sky. The Andersons figured that $350 million over a five-year period would be required to make their dream a reality. Their appeal reached the halls of the U.S. Congress, where Senator Alan Cranston of California lent his support for what could have been the world's first biomedical research initiative to come close to matching the size and scale of the U.S. Apollo space initiatives.

The Argonne group's cancer results, the culmination of nearly a decade of work, could have been interpreted as proof of the principles behind the Human Protein Index. Instead, most scientists took little or no notice of their report, which was published in 1984 in the rather obscure journal *Clinical Chemistry*.[4] Anderson and Taylor did not receive large allocations of new research funds, nor were they bestowed with awards. And why should they? The Argonne group certainly hadn't cured cancer. They hadn't even classified real human tumors. Instead, they used cultured cells derived from tumors and they used only a small number of samples, rather than larger and more statistically meaningful quantities. They hadn't shown that the categories in which they sorted their tumor cell samples were particularly meaningful. They hadn't correlated clinical outcomes or treatment responses with their computer-generated categories.

Indeed, the Argonne team appeared to be more interested in fundamental biological questions than in medical applications. They wrote, "Ideally, one would like to use a method that could, by itself, discover the underlying logical structure of the gene expression control mechanisms."[5] They felt that by electronically tracking protein changes in cells at various stages of development, one could deduce an underlying molecular "circuitry." Thus, the Andersons and their coworkers believed that they were onto a means of solving one of biology's most difficult riddles. How is it that one cell can give rise to so many different cell types, each containing the very same complement of genetic material? How does a fertilized egg cell differentiate into hundreds of specialized cell types, each appearing in precise spatial and temporal order? But these lofty scientific goals also garnered scant attention for the molecular astronomers, in part because the proteins were identified solely by position on the gel. The Andersons and their colleagues couldn't readily reveal their structures or functions. (This would require purification and sequencing of each protein spot, a prohibitively expensive and time-consuming task at that time.) It was hard to imagine the development of a scientific explanation for cellular phenomena that did not include knowledge of the structure and function of the relevant molecular components. Similarly, it was hard to imagine any physician being comfortable making a diagnosis based on a pattern of unidentified spots that was not linked to some plausible explanation. Furthermore, despite the Andersons' and their colleagues' best efforts, at that time two-dimensional protein gels were still difficult to reproduce in a way that would allow surefire alignment of identical proteins across gels. In any case, in the mid-1980s too many scientists felt that protein analysis technologies were still unwieldy, and too few scientists were compelled by the Andersons' vision of the future, so the Human Protein Index fell by the wayside. Thus, instead of being a catalyst for biomedicine's moon shot, the Argonne team's cancer work appears as little more than a historical footnote, or so it may appear.

When asked about these rudimentary experiments 16 years later, Leigh Anderson would have absolutely nothing to say. Was he discouraged by lack of progress or by years of disinterest by his peers? Hardly! The Andersons had managed to start a company back in 1985, aptly named Large Scale Biology Inc., and after years of barely scraping by, the Maryland-based company was finally going public. In the year 2000 investors had discovered the Andersons' obscure branch of biotechnology in a big way, and Leigh Anderson's silence was due to the self-imposed "quiet period" that helps protect initial public offerings (IPOs) from investor lawsuits. Leigh Anderson, Taylor, and a few dozen other research teams had made steady progress and, as will be shown in later chapters, the Argonne work from the 1980s was indeed very relevant to both medical applications and understanding the fundamental nature of life.

For the Andersons in 2000 the slow pendulum that carries the spotlight of scientific interest had completed a circle. It began for Norman Anderson in 1959, while at the Oak Ridge National Laboratory in Tennessee, where he first conceived of a plan to identify and characterize all the molecular constituents of human cells and where he began inventing centrifuges and other laboratory instruments useful in separating the molecules of life. The Human Protein Index was a logical next step. "Only 300 to 1000 human proteins have been characterized in any reasonable detail—which is just a few percent of the number there. The alchemists knew a larger fraction of the atomic table."[6] In other words, how can we build a scientific understanding of life processes or create rational treatments for dysfunctional processes without first having a catalogue or list of the molecular components of life? Imagine having your car being worked on by a mechanic who is, at most, slightly familiar with only 1 or 2 percent of the car's parts.

The Andersons' early 1980s campaign, their efforts to rally scientists and science administrators for a huge bioscience initiative, their call for a "parts list of man" with computer power to support its distribution and analysis, and their daring in laying forth their dreams . . . all of these did not vanish without a trace. They were echoed a few years later when scientists began to seriously contemplate making a list of all human genes and all DNA sequences. This led to the launch of biomedicine's first true moon shot, the Human Genome Project, and, leaping forward, to a December 1999 press release announcing that the DNA sequence of the first entire human chromosome was complete. The accompanying report, which appeared in *Nature* magazine, contained a treasure trove of information for biomedical researchers and served to remind the public that the $3 billion, 15-year Human Genome Project was nearing its end a full four years ahead of schedule. The entire DNA sequence of all 24 distinct human chromosomes, along with data on all human genes (and proteins), would soon be available.[7] In response, investors poured billions of dollars into companies poised to apply this new resource, including a few hundred million dollars for the Andersons' Large Scale Biology outfit.

As far back as the early 1980s, Leigh and Norman Anderson had contemplated what they referred to as a "list-based biology."[8] They had a vision of an electronic catalogue of the molecular components of living cells and mathematical analyses that would make use of this data. They had even gone so far as to suggest that a "list-based biology, which [the proposed Human Protein Index] makes possible will be a science in itself."[9] The Argonne group's cancer study, despite the fact that the proteins were identified only by position, was a prototype for this new type of biology. Many more would follow.

The search for a cure for cancer played an even bigger role in another landmark information-intensive research effort begun in the 1980s. It was initiated by the world's biggest supporter of cancer research, the U.S. National Cancer Institute (NCI). One of the NCI's charges is to facilitate the development of safer and more effective cancer drugs, and in the mid-1980s Michael Boyd and other NCI researchers devised an anticancer drug-screening initiative that was fittingly grand. About 10,000 unique chemical entities per year were to be tested on a panel of 60 different tumor cell cultures.[10] Each chemical compound would be applied over a range of concentrations. Each tumor cell culture would be assayed at defined time points for both growth inhibition and cell death.

Drug discovery has always been a matter of trial and error, and as medicinal chemists and molecular biologists became adept at synthesizing and purifying new compounds, preliminary testing became a bottleneck in the drug development pipeline. Laboratories from throughout the world would gladly submit compounds to the NCI for testing. At that time, researchers were looking for compounds that gave favorable cellular response profiles, and they were looking to further define those profiles. The NCI initiative would establish a response profile based on the pattern of growth inhibition and cell death among 60 carefully selected tumor cell cultures. A poison that killed all of the cell types at a particular concentration would not be very interesting, for it would likely be toxic to normal cells as well. However, compounds that killed particular types of tumor cells, while sparing others, could be considered good candidates for further studies. The response profiles of both approved cancer drugs and those that failed in clinical testing would be used as guideposts for testing the new compounds, and as new compounds reached drug approval and others failed, retrospective studies could be used to further refine model response profiles.

The NCI's bold initiative, named the Development Therapeutics Program, was launched in 1990, and by 1993 30,000 compounds had been tested on each of the 60 cell cultures. This work generated over a million points of data, information that had to be digitized and stored on a computer. How else could one build even the most rudimentary understanding of the actions of tens of thousands of different molecules? How else could this information be stored and shared?

The complexities of cancer placed tremendous demands on biologists and medical researchers—demands that could only be met through electronics and computation. The NCI's Development Therapeutics Program, like the Argonne protein studies, required machines to collect information and transduce it into electrical signals. There are many other ways of collecting data: For example, armies of technicians could take measurements by eye and record them in volumes of laboratory notebooks. However, as any citizen of the Information Age knows, for gathering, storing, and manipulating information, nothing beats the speed, cost, versatility, and ease of electronics. Information technology thus greatly facilitated the development of new efforts to understand and attack cancer. The NCI Development Therapeutics Program developed automated devices to read cell density before and after treatment. An electronic growth response curve was generated and for each compound the concentration responsible for 50 percent growth inhibition was automatically calculated. COMPARE, a computer program written by Kenneth Paull and colleagues at the NCI, compared response profiles, ranked compounds on the basis of differential growth inhibition, and graphically displayed the results.

Initially, the NCI's Development Therapeutics Program was only slightly more visible than the Argonne team's early protein studies. However, both initiatives shared a characteristic common to many information-intensive projects. Their utility skyrocketed after incremental increases in data passed some ill-defined threshold and upon the development of so-called "killer applications," computer algorithms that greatly empower users. By 1996 60,000 compounds had been tested in the Development Therapeutics Program and at least five compounds, which had been assessed in the screen and analyzed by COMPARE, had made it into clinical trials. New databases were then linked to the cell response data set, including a database of the chemical compounds' three-dimensional structures, a database of the compounds' molecular targets, and a database of the pattern of proteins that appear in the 60 targeted tumor cell cultures. With this last data set some of Anderson's two-dimensional gel electrophoresis work became enjoined with the work of the Development Therapeutics Program.[11] This data was all electronically linked and led to numerous scientific articles, including a prominent piece written by John Weinstein of the NCI, appearing in *Science* magazine in 1997.[12] In this outstanding article Weinstein and coworkers caution that "it remains to be seen how effective this information-intensive strategy will be at generating new clinically active agents."[13] Skepticism is a trademark of good science, but no one could possibly suppress the wonderment and pride that arises from this account of hundreds of millions of individual experiments distilled by powerful number-crunching algorithms and vivid color graphics into meaningful new medical leads and biological insights.

Is even one decisive victory over at least one major type of human cancer imminent? Four years after Weinstein's paper, the answer is still not clear. It *is* clear, however, that information technology has enabled fantastic new tools for unraveling the complexities of cancer. Some of these tools, their application to cancer and other disorders, and the prospects for new treatments will be discussed in greater detail in later chapters, as will the profoundly relevant and enormously information-intensive efforts to understand and account for our genetic makeup.

Does an information-intensive approach represent a revolutionary new framework for understanding cancer and other biological phenomena? Absolutely, for it prompts us look at life in a dramatically new way.

2

Information and Life

All kinds of electromagnetic emissions, from visible starlight to stealthy cosmic rays, flow through the heavens. But one particular category of emissions, those emanating from a nonterrestrial source and having a bandwidth of less than 300 Hz, has never been detected. Known cosmic phenomena, such as the fusion infernos within stars, exploding stars, black holes, and the big bang do not appear to release energy of such limited spectra. It is for this very reason that a group of radiotelescopes have been programmed to scan the universe in search for such signals. The radiotelescopes are under direction of the renowned SETI program, which for those that do not know, is a code name for the *Search for ExtraTerrestrial Intelligence*. The SETI team, which is composed of a group of scientists of diverse expertise, speculates that intelligent life might produce such a narrow-banded signal and that searching for such a signal represents one of humankind's best hopes of finding evidence of distant life. Of course, confirmation of extraterrestrial life would require more than just finding a narrow bandwidth signal. The hope is that a narrow bandwidth signal will be a carrier signal within which coded messages could be found. By focusing on such a signal with very sensitive instruments, such coded information might be detected. Coded information, irrespective of content, is indicative of life. Coded information hurled through space in the form of electromagnetic waves, it is presumed, would be indicative of intelligent life.[1]

Although most biology textbooks neglect to mention it, information is as fundamental and unique to life as either metabolism or reproduction. Encoded messages occur in a myriad of forms and are transmitted between a myriad of different types of receivers and senders. Information is sent when a beaver slaps its tail on the water upon sensing a danger, when a plant releases a fragrant odor, when a bacterial protein signals a gene to turn on the production of catabolic enzymes, and when a nerve impulse causes a muscle to contract. In each case, whether between organisms or within an organism, a coded message is provided. Communications such as these are not just the foundation of life, they are its essence.

Human beings have distinguished themselves among other species on earth by continually developing and adopting new and improved ways of exchanging information. No other species comes close to matching our language, speech, and writing capabilities. Of course, human ingenuity and innovation have not stopped there. Modern information technology greatly facilitates the storage, processing, and conveyance of information. Weightless or nearly weightless electrons and electromagnetic waves travel at or near the speed of light (almost one billion feet per second). And, after only a few hundred years of development, these information conduits have enabled stunning advances, such as the Internet and other global communication networks and a machine that can outperform the best living chess player.

Welcome to the Information Age, where the movement of speedy electrons and electromagnetic waves has replaced much of our mechanical and mental work. Fewer and fewer people turn knobs on TV sets, rotate dials on telephones, write letters by hand, and/or tally bills on abacuses. We can (and, more often than not, want to) do it faster, better, and cheaper using devices that transduce our thoughts or desires into electronic signals. We telecommute and use e-mail, remote control devices, voice recognition software, and so forth. The resulting electric signals may be readily digitized, processed, stored, replicated countless times over, and transmitted over long distances before being converted into images, sounds, or other stimuli suitable for the human senses.

Information circuitry is not only external. The senses are portals to an internal information network. Ears, eyes, nose, skin, and mouth convert patterns of touch, sound, odor, taste, or light into patterns of nerve impulses. This information is passed along by waves of ions (charged atoms or molecules) moving across the outer membrane of neurons (nerve cells). These waves (action potentials) move at rates of up to 100 feet per second (68 mph). This may seem pitifully slow compared to today's electronic and electromagnetic speedways, but at the time that it developed, beginning about 500 million years ago, neuronal signaling was revolutionary. Neurons provide a much faster and more efficient channel of communication than either chemical diffusion or any sort of fluid pump. Neuronal signaling allowed quick and coordinated responses to environmental stimuli. It also led to the development of the brain, cognition, and consciousness.

The nervous system was derived from and superimposed upon yet another information network, a more ancient network, and the mother of all information networks. Living tissue is composed of millions of different proteins, nucleic acids, fats, and other chemical entities. These are the molecules of life and the subject of countless research studies. They can be understood in terms of their physical structures and their chemical properties. However, they can also be understood in terms of the information that they convey.

What kind of molecular messages are being sent within us? Who is the sender and receiver and what is being said? Genes are the most obvious conveyors of information within living beings. Gregor Mendel first characterized genes as units of hereditary information, agents responsible for determining particular traits and characteristics that are passed from parent to offspring. He referred to them as *factors* in an 1865 publication, but at that time and for many decades thereafter no one knew precisely how these units of hereditary information were stored or how they translated into traits and characteristics. In 1943 Erwin Schrödinger speculated that genes were "some kind of code-script," and this is indeed the case. We now know that genes are encoded by a series of molecules known as nucleotides, which are the components of deoxyribonucleic acid (DNA). Genes provide coded instructions for the production of millions of additional nucleic acids and proteins.

The word *factor* has a physical connotation, and thus from the very start genes have had dual personalities. Like a photon of light teetering between matter and energy, a gene teeters between matter and information. On one hand, a gene is not unlike the 100 or so chemical elements of nature. Each gene may be a distinct and rigidly defined composition of nitrogen, oxygen, hydrogen, carbon, and phosphorous; one that participates in a series of chemical reactions that results in the production of additional nucleic acids and proteins. But this is like saying that the United States Constitution is a particular construction of plant pulp and ink. Genes encoded in DNA convey information to additional nucleic acids, which relay the messages to proteins, which convey signals throughout the organism. A complete set of genes, the genome, carries instructions, or a blueprint, for the development and function of an entire organism.

As information it does not really matter how the gene is encoded, so long as the message can be received and decoded. There is redundancy in the genetic code; the same gene may be encoded by any one of a number of different nucleotide sequences. A gene may also be encoded in an entirely different medium. In a classic instructional film from the early 1970s, a DNA sequence is portrayed by dancers, each wearing one of four brightly colored costumes, representing four types of nucleotides. The dancers simulate the production of a protein through their choreographed movements. Nowadays, no matter whether the gene is encoded by a string of nucleotides, costumed dancers, words, or 0s and 1s (binary code), in a laboratory it can be readily converted into biologically active proteins.

Other types of molecules also convey messages. For example, growth hormone, a protein, carries a message from the pituitary gland in an animal's brain to muscle and bone cells throughout its body. Again, the message is encoded in molecular structure. A series of molecular subunits tell the molecule to assume a highly specific and otherwise improbable shape. Receptor proteins on the surface of muscle and bone cells recognize the structure of the hormone. They read the message, process the information, and pass on their own signal to other molecules within the cell. These molecules do the same, and eventually this network of signals results in cell growth division. The growth hormone molecule has thereby told the cells to grow and divide.

All the molecules of life can be thought of as information carriers. The implications of this are profound. The pattern of proteins that a cell produces is not just information about the cell; it is a message from the cell. Proteins communicate among themselves and with other molecules within the cell. Those on the cell surface and those secreted by the cell are messages sent to other cells. The Argonne group intercepted some of this information when they broke open cells and captured their individual proteins in spots on a gel. Gene sequencing captures a different set of molecular messages than does two-dimensional gel electrophoresis, and new technologies may intercept many more of these internal communications. Molecular messages may move quite a bit slower than the speed of light, but as information they are essentially no different than messages sent over phone lines or reflected off of satellites: they all may be transduced, digitized, and stored. Having molecular information in digital format brings the study of life to an entirely new level, because it facilitates analytic techniques that rely on and exploit the mathematics of probability. Why probability? Because mathematical odds are at the foundation of information science.

In its most common usage, information is news. It may be a weather report, a football score, a department store sale, a sibling's birthday, a bank account statement, a competitor's market share, or the price of tea in China. In the living world, information flows like water from a ruptured fire hydrant. In today's Information Age it races through information superhighways—global networks of phone lines, wires, satellites, faxes, cell phones, pagers, computers, and the like. It is in this realm, the electromagnetic one, that a mathematical theory of information first emerged.

The invention of the telephone provided a means of transmitting sound through copper wires. However, making out what someone is saying over a phone line, particularly at great distance, has always been a problem. Not only is it difficult to precisely replicate the sound of the voice entering the receiver, but noise tends to enter the system. Ma Bell, the principal U.S. telephone company throughout most of the twentieth century, had an interest in the advancement of phone service and in determining the relative costs of phone calls that could vary in duration, distance, and message content. It was within the phone company's research labs that Claude Shannon, a research mathematician, first developed mathematical equations for describing how information is communicated. His landmark paper, "A Mathematical Theory of Communication," appeared in the *Bell System Technical Journal* in 1948 and is considered to be a foundation for the science of information theory. "The fundamental problem of communication," wrote Shannon, "is that of reproducing at one point either exactly or approximately a message selected at another point." In order to describe this problem from the vantage point of the signal carrier, in this case the copper wire, he stripped information of any meaning. "Semantic aspects of communication are irrelevant to the engineering problem," he wrote.[2] Thus, a telemarketing scam and a declaration of war may have equivalent information content.

Shannon defined information content in terms of its newness. A communication doesn't contain information if it restates what is known or what has already been anticipated. To be information a message must be somewhat unpredictable to the receiver. A predictable message, such as a repeating pattern of words or syllables, does not convey additional information. An entire book can be written with the words "biology is about communication," repeated 10,000 times. It would be mundane, boring, monotonous, dull, and altogether a tedious read. The same information can be transmitted in just a single sentence, as was just done. The content is the same. It has just been compressed so that each symbol brings something new. (Of course, the phrase "I love you," may convey new information each time it is said, for even without variations in intonation, emphasis, or pausing, it may have a different contextual meaning each time. Sequences of nucleotides may also be context-dependent, as will be discussed later).

To capture the newness idea, Shannon considered a message as a sequence of symbols (they could be 0s and 1s, letters, words, or anything else). He then developed a way to measure the chance that a symbol will be unanticipated. If there are N choices of symbols, each with an equivalent likelihood of occurring, then the probability that any one symbol occurring is 1 divided by N. The probability of any two independent events occurring is equal to the product of each single probability ($P_{12} = P_1 \times P_2$). To make information rise as the probability decreases, Shannon made information (I) proportional to $1/P$. Then, to make independent units of information additive (so that $I_{total} = I_1 + I_2 + I_3 \ldots$) Shannon took the logarithm of the probabilities, therefore Log ($1/P_{12}$) = Log ($1/P_1$) + Log ($1/P_2$). Shannon then defined the smallest piece of information as one *bit*. This occurs when there are two choices (such as 0 and 1), each occurring with the same overall frequency. P is therefore 2, and to make Log (2) = 1, base 2 is used; thus I = $\text{Log}_2(1/P)$ and the equation for the total information in a string of symbols is $I_{total} = \log_2(1/P_1) + \log_2(1/P_2) + \ldots$. Shannon's mathematics and subsequent work by Warren Weaver, Andrey Kolmogorov, and many others proved to be extremely useful not just for the communications industry, but also for an emerging computer industry, where the coding and storage of information is of fundamental importance. Thus, the concept of information became firmly grounded in the mathematics of probability.

The mathematical derivation of information content happens to parallel that of entropy, a measure of thermodynamic disorder.[3] The nineteenth-century physicist Ludwig Boltzmann determined that the entropy of a particular multicomponent structure (where a component is anything equal to or larger than an atom) is related to the number of arrangements that the components could have and the probability that they would happen to be arranged as they are. A pile of sand would have high degree of entropy (disorder), while a sand castle would have a low degree of entropy. Similarly, a string of random amino acids (the components of proteins) that assumes no particular shape would have a high degree of entropy relative to an intricately folded string of amino acids, shaped by evolutionary forces to fulfill a particular function or convey a particular message. Low entropy is analogous to high information content and vice versa. The only difference between Boltzmann's entropy equation and Shannon's information equation is that the entropy equation has a negative sign and its probabilities are multiplied by a constant (*Boltzmann's Constant*, which has units of calories per degree Centigrade). The significance of the similarities between entropy and information is the subject of fierce debate, and this area of thought appears to be fertile ground that may someday yield important new intellectual breakthroughs. Information science is important to the study of life. Probabilistic analyses not only underlie the search for extraterrestrial beings; they also underlie the search for terrestrial life's internal intelligence, the signals hidden within DNA, proteins, and other molecules of life. Where will these studies lead?

An information-centric view of life raises some intriguing possibilities. If *all* information from a living system can be captured and digitized without changing its fundamental nature, then it seems reasonable to conclude that the creation of electronic or *in silico* life may be possible. Science fiction enthusiasts have been quick to point out the possibility—or to some the inevitability—of such life. When one considers how quickly people have learned to faithfully transfer complex sounds and moving images into binary code, as well as the rapidly growing prevalence of increasingly sophisticated electronic aids, the prospect of some sort of leap into electronic life may seem closer with each new software release.

However, far more powerful and more immediate forces are working in the opposite direction. The desire to know ourselves and the even keener desire to simply maintain our flesh and blood structures are propelling the course of an enormous volume of biomedical research. The manifestation of these forces is described in the following historically oriented chapters.

3

Behold the Gene

Among the greatest scientific achievements to date has been the identification of various disease-causing and disease-associated genes. Knowledge of disease genes provides an explanation of how and why particular people fall ill. It brings us out of the Dark Ages.

In 1948 Linus Pauling initiated a new era in biomedical science when he discovered that a particular heritable variation in a single component of the hemoglobin protein complex results in sickle-cell anemia, an often-deadly blood disease. Pauling was able to distinguish the protein variant from others and causatively link it to the disease. It was an extraordinary triumph because it provided a single tangible origin for the fragile sickle-shaped red blood cells that are the hallmark of the disease. Pauling had inextricably linked a disease to a molecule, identifying a culprit that could be tracked from parent to offspring. Years later, the gene responsible for the hemoglobin variant was discovered, and the entire series of discrete molecular events leading to the symptoms of the disease was fully delineated. The gene dictates the protein sequence, which influences the shape and chemical properties of the hemoglobin complex, which in turn determine the shape of the red blood cells. The sickle-cell hemoglobin variant tends to give the oxygen-carrying sacks a sickle shape, causing them to burst upon passing through narrow capillaries, thus reducing the oxygen carrying capacity of the blood and bringing on the ill effects of anemia. The delineation of this pathway was particularly satisfying because it fulfilled a widely held scientific aesthetic: Not only did it demystify a natural phenomenon by reducing it to known matter operating under established laws, it left no loose ends. The explanation was complete and it rested on a single entity, the disease gene.

The scientific underpinnings of this triumph began in the garden of Gregor Mendel in 1865. There, the Czech monk found that particular traits in peas and other plants are the consequence of the action of sets of discrete factors that are transmitted from parent to offspring. The factors, now known as *genes*, come in various forms, known as *alleles*. One allele may be dominant over another, or each may contribute to the trait. Thus, the particular combination of alleles that an offspring inherits determines its *phenotype*, which is the set of traits that the offspring displays. Different genes may segregate independently. For example, a pea may inherit a green color from one parent and a crinkly surface from the other parent. Mendel's great achievement was to show that the pattern of inheritance follows a set of broadly applicable rules. His investigations went unrecognized for 34 years, but eventually spawned a mathematical framework (known as Mendelian genetics) for describing inheritance.

In 1902 Walter Sutton and Theodor Boveri recognized independently that the inheritance pattern of Mendel's factors paralleled the inheritance pattern of particular microscopic structures known as *chromosomes*, found within dividing cells. Working with fruit flies, Thomas Morgan, a Columbia University professor, and Alfred Sturtevant, a student in his lab, examined the frequency by which pairs of genes were cotransmitted from parent to offspring and by 1915 concluded that a linear arrangement of genes on a chromosome best explained the patterns of coinheritance. Although genes could not be directly observed, they could be genetically mapped, or ordered on a chromosome, by determining how often various pairs were inherited together. In the course of their work Morgan and coworkers identified many inherited fruit fly characteristics and attributed a distinct gene to each, thereby establishing the one gene–one trait postulate of genetics and prompting a century of chatter about "a gene for _____," where the blank could be almost any characteristic, behavior, or disease.[1]

Trait-causing factors in humans and other sexually reproducing organisms adhere to the same inheritance rules as those in the pea plants and flies that Mendel and Morgan studied. As early as 1907 eye color and the metabolic disease alkaptonuria were shown to follow Mendelian rules of inheritance. In the years that followed, Mendelian genetics would also be shown to determine the pattern of inheritance of numerous other traits. But the predictive powers of Mendelian genetics are limited. Most often, the rules predict the percentage of offspring that will inherit particular combinations of alleles, but not whether any particular individual will inherit any particular combination. Nor do Mendelian genetics explain the mechanism by which an inherited factor gives rise to a particular disease or trait.

The molecular genetics of Linus Pauling picked up where Mendelian genetics left off. Sickle-cell anemia was the first disease to be characterized at a genetic level. Many more would follow and to date more than 6500 gene variants have been linked to human phenotypes.[2] In most cases, knowledge of the molecular biology that underlies the disease process, often gleaned from animal studies, leads to the identification of the disease gene. For example, once researchers had delineated the molecular pathway by which sound is transduced into nerve signals, they were able to examine the genes from individuals with inherited forms of deafness and discover alleles of at least 30 genes that may be responsible for the deafness.[3]

Sometimes little or no knowledge of the disease process is required to find the responsible gene. If a sufficient number of large families with numerous affected members exists, a disease can be linked to a gene through chromosome mapping. Genes are contained in chromosomes that exist within each cell of the body. In humans a set of 23 chromosomes is inherited from each parent (see Fig. 3.1). By chemically treating cells and splattering them on a glass slide one can release their chromosomes, which can then be stained and observed under a microscope. In 1968 Roger Donahue, a graduate student at Johns Hopkins University, noted that one of his own chromosomes was unusually long. Staining revealed that a particular region of it was extended. Donahue collected blood samples from his extended family, examined their chromosomes and their blood proteins, and linked the inheritance of a particular protein variant (the protein responsible for the Duffy-a blood type) with the inheritance of the long chromo-

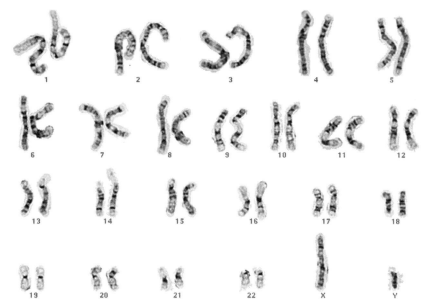

FIGURE 3.1 One complete set of human (male) chromosomes. *(Image courtesy of Xu Li, Kaiser Permanente Regional Cytogenetics Laboratory, San Jose, California.)*

some.[4] The gene for the blood protein was thereby physically mapped to a chromosomal region. The occurrence of the protein variant and the extended chromosome were perfectly correlated.

The occurrence of a disease can be correlated with the occurrence of one of many tiny chromosomal variations, just as the blood protein variant was correlated with the occurrence of an extended region of the Donahue family's large chromosome. Each mapped chromosomal variation becomes an additional marker for determining the relative position of new genes. As germ cells (sperm and eggs) are made, pairs of chromosomes exchange fragments (see Fig. 3.2). Over several generations two variations that are close to one another on the same chromosome may become separated. Over millions of generations so many have separated and rejoined that they become randomly mixed. When they are thoroughly mixed, they are said to be *in equilibrium*. Prior to this they are in what is known as *linkage disequilibrium*. Imagine a deck of cards arranged so that the ones precede the twos, which precede

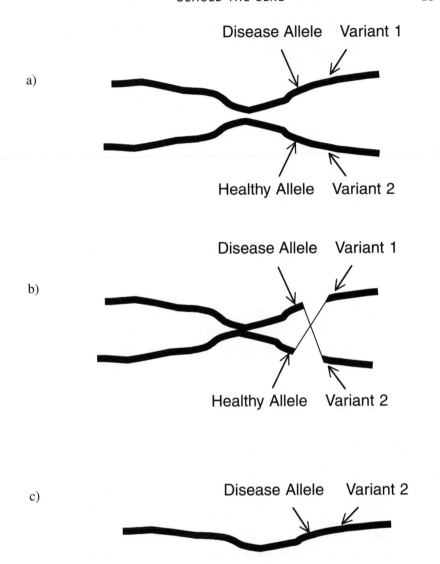

FIGURE 3.2 Genetic recombination (*a* through *c*) during egg or sperm cell development. Chromosome strands come together, break and rejoin, and then separate. Offspring will inherit either strand. The "Disease Allele" is now coinherited with gene "Variant 2" rather than gene "Variant 1."

the threes, and so on. After one shuffle, cards of one number still tend to be associated (linked). However, after a large number of shuffles the card order is random.

Linkage disequilibrium allows us to determine gene order. On average, the closer two variations are to each other the longer it will take for them to separate, and the more likely it is that they will be inherited together. The farther apart two sites are located on a chromosome, the greater the chance that a fragment exchange will separate them. Thus, the *recombination frequency*, a measure of how frequently a disease and a particular marker no longer co-occur, is related to the physical distance between the underlying disease gene and the marker.

The recombination frequency can therefore be used to place the gene on a chromosome map. Knowledge of a gene's approximate chromosomal location can often help geneticists identify and isolate the gene. This procedure, known as *positional cloning*, is far from straightforward, for genes may occupy less than 0.001 percent of the identified chromosomal region and may be difficult to discern. The region can often be narrowed by using newly discovered markers and by testing more affected families. Having more markers is like having more landmarks on a map, and having more affected family members allows a more precise measurement of genetic distance (recombination frequency). Ultimately, a variety of additional techniques must be employed to ferret out and test candidate genes within the stretch of DNA. Through various combinations of skill, intuition, perseverance, and luck, over a hundred of these needles in the haystack have been found.[5] Genes have been identified for Huntington's disease, cystic fibrosis, breast cancer, colon cancer, Alzheimer's disease, and Duchenne muscular dystrophy among others. Positional cloning has also led to a gene that is responsible for the very common affliction known as maleness. SRY (for *sex* determining *region, Y* chromosome) has been shown to induce the formation of testis, which in turn prompts the development of most male features.[6] Individuals lacking SRY appear female, whether or not they have other portions of the Y chromosome.

The gene provides a starting point for the delineation of the downstream events, our development. Additionally, with the gene in hand, one can study the source of the mutational event from which a heritable disease gene emerged and speculate as to how it may have originated and why it has persisted. The sickle-cell allele of the beta-globin protein is most frequently found among people whose ancestors lived in areas with a high incidence of malaria, such as equatorial Africa. The allele is thought to have arisen from a mutation in an ancestral gene that occurred on several occasions millions of years ago. This mutation persisted because it conferred an evolutionary advantage: In the heterozygous state (inherited from only one parent and thus combined with the normal allele from the other parent) it protects the individual from malaria. Only when it is inherited from both parents does it cause sickle-cell anemia.

When any new disease-causing entity is found, it can whip up a frenzy among health care workers analogous to the frenzies precipitated by uncooperative dictators or terrorist groups. Researchers, clinicians, doctors, nurses, and others have rallied against HIV (the virus that causes AIDS), helicobactor pylori (the bacteria that induces ulcers and stomach cancer), malaria bug, the smallpox and polio viruses, and the like. The progress in each of these wars has been remarkable. The discovery of a disease-causing gene incites a similar reaction, touching off yet a new offensive in the battle against disease. First, the identification of a disease gene enables highly accurate predictions to be made regarding whom will be afflicted. For example, soon after the Huntington's Disease Collaborative Research Group discovered the gene responsible for Huntington's disease, a diagnostic test was developed. Huntington's disease (also known as Huntington's chorea) is characterized by dementia, rigidity, seizures, and spasmodic movements that typically begin between the ages of 30 and 40 and end with death about 17 years later. Prenatal testing for Huntington's disease is now common and is affecting people's decisions to have children. Second, a defective gene may in theory be replaced. If the nondisease allele can replace the disease allele, a cure may result. Gene therapies, any of a variety of experimental treatments that involve administering genetic material, are slowly advancing. The difficulty lies in the fact that the disease allele lies in every cell of the body. Delivering the genes to the appropriate cells, persuading them to stay there, and getting them to express the appropriate amount of the missing protein at the appropriate times are all formidable technological hurdles. Nonetheless, progress is being made. Cystic fibrosis patients suffer from chronic bacterial infections in their lungs due to a mutation in a protein that normally acts in the elimination of such bacteria. Hundreds of cystic fibrosis patients have undergone experimental gene therapies. Despite the lack of an overwhelming success, these clinical trials have provided at least temporary relief for some patients and have motivated many researchers to further pursue such treatments.[7] Third, the identification of disease genes immediately opens up new avenues of research that lead to additional insights into the disease processes. These in turn may lead to better therapeutics. The identification of disease-causing genes has made a significant impact on human health and has profoundly enriched our hopes for even greater advances in human health. Scientific findings not only explain and predict things; they also provide solutions to people's problems, or in economic terms, goods and services to meet our needs and desires. Therefore, it is of no surprise that commercial interests would play a role in heralding gene discoveries as crowning triumphs.

In 1994, an eight-year positional cloning effort by researchers at Rockefeller University culminated in the identification of a gene for a protein that was christened *leptin*, from the Greek word for thin.[8] Leptin mutations in mice result in obesity, and the administration of the leptin protein to either the mutant mice or to normal mice results in a decrease in appetite, an increase in energy expenditures, and consequentially in weight loss. Amgen Inc. promptly paid Rockefeller University $20 million and promised royalties of up to $50 million for exclusive rights to develop weight loss products from the human version of the leptin gene. The announcement reverberated through the investment community. Leptin was very new, little was known about the human version of it, and no human disease was known to be associated with the gene. Yet clearly expectations were sky-high. Gene patent protection had already been established, and the gene products tissue plasminogen activator (a blood-clot buster), human insulin (for diabetes), and erythropoietin (to replenish blood cell growth after dialysis or cancer therapy) had already each garnered hundreds of millions of dollars in drug sales.

Consider all the things that single genes have been shown to do. One gene may provide the cornerstone of an explanation of a previously inexplicable human condition, be used to diagnose a disease, effectively treat a disease, or yield its discoverers millions of dollars. A single molecule may raise the hopes, spark the imaginations, and inflame the passions of patients, health care workers, and all sorts of investors. Not surprisingly, in the realm of the life sciences the gene has been crowned king. Around it has been constructed a framework for biological research known as *genetic determinism*. It is a scientific paradigm that has guided legions of researchers in pursuit of master regulatory molecules, disease-causing genes, and genes responsible for every imaginable trait or condition.

But the one gene–one trait postulate can be pushed too far. Often research findings are distorted to meet the expectations of the paradigm, obscuring a harsh reality: that many genes have multiple functions, that diseases and traits often defy single-gene explanations, that the whole process of searching for the deepest causes of disease is beset with mind-numbing complexities, and that drug discovery is still ruled by serendipity. These issues will be examined later in the context of a revolution in our understanding of living organisms, one necessitated or elicited by this scientific gambit.

4

Working with Genes
One at a Time

It is quite common nowadays for high school and college students to purify human DNA (deoxyribonucleic acid derived from human cells) as part of an introductory biology lab course. They precipitate the DNA in a beaker of ethanol, capture the white viscous stringy material on a toothpick or glass rod, and then ponder the fact that what they are looking at contains instructions that define many of their physical, mental, and emotional characteristics. DNA is a polymer, a class of molecules that are composed of building blocks (monomers) linked together in a chain. Polymers constitute much of the material of living organisms. They are the molecules of life. Proteins, nucleic acids (DNA and RNA), and fats are chains of amino acids, nucleotides, and fatty acids, respectively. Like strings of any finite group of units, such as letters or words, they can be constructed of an infinite number of different sequences, each of which may have distinct properties (see Fig. 4.1).

Chromosomes are composed of both nucleic acid and protein matter. A series of experiments in the 1940s and 1950s led to the discovery that genes are physically embodied within the nucleic acid component. In the "blender experiments" of 1952, Alfred Hershey and Martha Chase tagged both nucleic acid and protein particles in a virus and showed that only the nucleic acid transmitted the viral genetic information. The particular type of nucleic acid was DNA, which is composed of the monomers adenine (A), thymine (T), cytosine (C), and guanine (G), linked by a backbone of repeating sugar and phosphate molecules. In 1953 James Watson and Francis Crick determined that DNA is a double helix, two strands oriented in opposite directions and

Monomer	Polymer
Amino Acids: Cysteine (Cys), Alanine (Ala), Proline (Pro)...	Protein: Met-Cys-Gly-Pro-Pro-Arg...
Nucleotides: Adenine (A), Cytosine (C), Thymine (T)...	DNA: ACTGGTAGCCTTAGA...
Letters: A, B, C...	Words: CAT, GO, FRIEND ...
Symbols: 0,1	Binary Code: 1001011100101...

FIGURE 4.1 Monomers join together to form polymers. Their sequence conveys information.

twisted around each other. The As on one strand align with the Ts on the other, and the Gs pair with the Cs, thereby forming bridges between the two strands like rungs on a ladder. One strand is complementary to the other. If the two strands were to come apart, each strand could then be re-created from the other strand, for each strand can act as a template. This is in fact what happens during DNA replication, enabling gene copies to be passed from mother cells to daughter cells, including the reproductive cells from which new life emerges. Such are the molecular actions that underlie biological inheritance. (The pairing of As with Ts, and Cs with Gs, is particular noteworthy, because it also underlies some of the most important technology used to study and modify living organisms.)

Proteins are the agents that drive most of the chemical reactions that animate living things, and in 1957 Crick and George Gamov determined how DNA directs their production. According to the "sequence hypothesis," the particular order of A, T, C, and G monomers (the DNA sequence) specifies the amino acid sequence of proteins. Crick and Gamov also suggested that genetic information flows only in one direction, from DNA to RNA (ribonucleic acid) to protein. Crick explained that "information means here the precise determination of sequence." With the exception of some viruses, such as HIV, the DNA to RNA to protein flow chart appears to be universal. It is known as the central dogma of molecular biology. The beta globin gene, for example, lies along a stretch of DNA that spans about 2000 base pairs. RNA is transcribed from the DNA in accordance with base-pairing rules. Portions of the RNA are removed, and the remaining RNA is modified slightly. The resulting RNA, known as *messenger RNA* (mRNA, also known as a *transcript*) is translated into a protein of 147 amino acids in accordance with a set of rules known as the *genetic code*. In this way the gene is said to be "expressed" (i.e., used to direct the production of mRNA and/or protein).

In 1966 Marshall Nirenberg, Heinrich Mathaei, and Severo Ochoa demonstrated that sequences of three nucleotide bases (codons or triplets) determine each of the 20 amino acids. In addition, particular triplets determine the beginning and end of protein synthesis. There are 64 combinations of three that can be derived from the four nucleotides. Nirenberg, Mathaei, Ochoa, and other researchers identified the triplet(s) that code for each the 20 amino acids, as well as the stop and start codons. The sickle-cell anemia allele of the beta-globin gene was found to differ from the more common allele by a single nucleotide. This nucleotide difference results in a difference in a single amino acid, which results in the difference in the hemoglobin protein, and so forth.

With the discovery of the nature of DNA and the genetic code, a gateway was opened up into a previously unknown world. It was an entryway into a rather special territory, one that lies not in depths of the seas or beyond the earth's atmosphere, but within each of us, within each of our ten trillion or so cells, and within the cells of all other living creatures. Insights into everything that was considered to be human nature, from disease propensities to dispositions, were poised to enter humankind's collective consciousness. And that's not all. "Evolution is an enchanted loom of shuttling DNA codes, whose evanescent patterns, as they dance their partners through geological deep time, weave a massive database of ancestral wisdom, a digitally coded description ancestral worlds and what it took to survive in them," the evolutionary biologist and writer Richard Dawkins has eloquently proclaimed.[1] Access to this ancestral wisdom would provide humankind with a better explanation of human origins, of the relationship between humankind and other forms of life, and of the relationship between individuals within our species.

Gene sequencing technologies, the means of determining the sequence of nucleotides in a given strand of DNA, were first developed in the 1970s by Fred Sanger, Allan Maxam, and Walter Gilbert. In one procedure, copies of a double-stranded fragment of DNA were split apart and DNA synthesis was made to occur on single strands (the templates). Radioactive (or later fluorescent) particles were incorporated into the new strands, then synthesis was terminated by the incorporation of crippled nucleotides (ones that do not support further growth of the chain) that had been spiked into the reaction mix. There were four different reactions in four different tubes, each with a different crippled nucleotide. The radioactive products in the tube with the crippled As were of varying sizes, but each ended in an A. The radioactive strands in the other tubes were of a different set of sizes and ended with either a C, T, or G. The synthesized products of each tube were separated by an electric field such that strands that differed by only one nucleotide could be distinguished. The four sets of reaction products were run adjacent to each other and yielded a ladder of radioactive bands arranged in a distinct pattern. Typically, these sequencing ladders were read by eye, and letters representing the bases were typed onto a computer file. By the 1980s, sequencing became routine for molecular biology labs. The technique was somewhat tricky and error-prone, but within the grasp of most technicians and students (see Fig. 4.2). As many as 1000 base pairs could be determined in one sequencing reaction, and by 1990 a total of 35 million base pairs of DNA had been sequenced.[2]

Throughout the 1970s, 1980s, and 1990s, scientists everywhere flocked to molecular studies. With tremendous vigor we identified and examined individual genes, their DNA sequences, and the RNA and proteins that were derived from them. Not only were various diseases distilled to the actions of individual genes, but so too were virtually all biological processes. DNA replication; cell division; evolutionary change; biological clocks; yeast mating; fruit fly development; transplant organ rejection; the regulation of

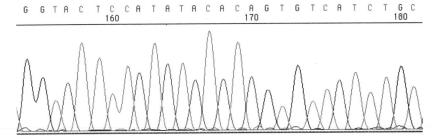

FIGURE 4.2   Readout from an automated sequencing machine. The horizontal axis represents the length of the DNA fragments (position in a sequencing gel). The vertical axis represents the intensity of four fluorescent dyes (one for each nucleotide). *(Image courtesy of Yiding Wang, Biotech Core Inc., Palo Alto, California.)*

blood sugar levels, body weight, and blood pressure; the production of nicotine in tobacco, of fat in cow's milk, and so on were all described in terms of the genes involved. Genes were overexpressed, underexpressed, eliminated, or mutated, and their effect on cells, organisms, and other genes was monitored in a variety of ways under a variety of conditions. Gradually, hierarchies of gene interactions were determined. Gene X produces a protein known as a *ligand*, which stimulates a complex of proteins known as a *receptor*, which activates a particular gene, which in turn does something else.

Most of the scientists who gravitated toward molecular studies set out to answer specific biological questions (such as how are nerve connections made, or why do plaques form in arteries?), but very often they spent many years studying particular genes.

Imagine being surrounded by darkness. You grope around and then trip over something, pick it up, and learn something about it. Suppose it is a chair. You grope around some more, and slowly you become familiar with the objects around you and their relationships to each other. You find a desk and a bookshelf with a few books on it. Perhaps you think that you are in a tidy little room, but then suppose you stumble onto another chair, a desk, and then a few more chairs. A furniture store? Okay—but wait, you just stumbled upon something else amid the darkness. What is it? It appears to be fabric. Then you discover some pillows, then a whole bunch of little boxes with liquid-filled plastic tubes in them, and then finally you stumble onto a large contraption, which after a considerable amount of time you determine to be some sort of barbecue grill. It turns out that you are in a huge and unfamiliar department store. You are in a foreign land, and many of the objects in the store are unknown to you. There is a design to it all, and you could in time become familiar with the contents of the store, and their relationships to each other, and the layout of the place. You may first develop in your mind a skeleton-like map of the layout. Gradually, as you explore, you may expand this map and fill it in with more details. As more and more details emerge, your mind may struggle to integrate and keep track of all that you have found. Collectively, this is what molecular biologists have been doing with the molecules in living organisms (although the objects of study are more often thought of as components of an intricate machine than as items in a dark department store).

Organisms are densely packed with molecules, and we have been tripping over them slowly, one by one. For example, decades ago cholesterol was found to play a role in the development of artery-narrowing plaques. Over the years, cholesterol studies led to the discovery of a few dozen genes involved in cholesterol production, regulation, or transport. They also led to a few dozen genes involved in hormone synthesis, reproduction, or a variety of other functions. A molecular understanding of cholesterol regulation and its role in atherosclerosis has been unfolding for some time, but there is still more to know. New molecular interactions continue to be found, and, despite decades of research, new components of the system continue to be stumbled upon. What about the researchers who are studying these molecules? Each new gene can be the basis for years of study and many doctoral theses. A group of students and other scientists may work on a set of closely associated genes in a single laboratory that is under the management of one gene expert. Their gene discoveries, particularly ones involving disease genes or other very important genes, may provide the basis for grant proposals and the initiation of their own independent labs. Thus, the social structure of biologists and their labs may recapitulate the structure of the genome. Science is, of course, a social endeavor and forces that affect the social interactions of scientists affect the course of science. Political pressures during the time of Stalin's reign, for example, have been blamed for setting Soviet biology programs on a course that hindered scientific progress for decades thereafter. Nowadays, quasi-socialistic institutions (groups that are far better at channeling the collective will than any Soviet government ever had) back the vast majority of life science studies. Governments, the U.S. in particular, and nonprofit foundations fund the majority of life science research labs. Most of these labs are housed in educational institutions, which usually pay the salaries of the lead investigators. These funding sources support a fiefdom of research laboratories, many studying particular molecules or groups of molecules. They also enable and enforce a code of conduct that plays a vital role in the advancement of biological knowledge and health care. Generally, the funding sources insist that the output of publicly funded science be made public and available for additional nonprofit work. Moreover, they actively seek to advance the means of sharing research findings, as will be discussed in the following chapter.

The sharing of research findings between two scientists may seem as placid as the bleat of a lamb, but as more sources of information become more effectively joined, the collective power of this information is more akin to the thunderous roar of a herd of buffalo. Biological information is a mighty beast, one that with the help of capitalist forces and the actions of those outside the mainstream of life science research has been unleashed from its pen. Some cowboys and cowgirls have jumped on it for the ride of their lives. Others have sought to put the beast back into its pen or to find other means to contain it, and many more look at it with a wary eye. Biological information and the forces shaping it and trying to direct it will be considered in coming chapters. And the white viscous material that dangles at the hands of students, too, will be further considered.

5

The Database

About two millennia ago, Alexandria, an Egyptian port city lying near where the Nile River meets the southeastern portion of the Mediterranean Sea, was a flourishing center for both commerce and scholarship. It was a key source of writing materials (papyrus) and a central hub for trade with numerous African cities, European ports, and various gateways to the Middle East and beyond. At the heart of this activity was the Alexandrian Museum and Library, a meeting place for scholars from different cultures and a repository of ideas, in the form of tens of thousands of historical and contemporary texts from Greek, Assyrian, Persian, Egyptian, Jewish, Indian, and other sources; works that were printed on papyrus or on parchment skins, rolled into scrolls or bound into books, housed in rows of chests and shelves, and catalogued by subject. This intellectual hub brought forth an enormous concentration of important discoveries and inventions. The concept of latitude and longitude, charts of constellations and stars, and more accurate time measurements aided in navigation and map making. A year was found to be 365¼ days, and the modern calendar was created. The size and orbits of the Sun and Moon were found, the circumference of the earth was calculated, sea routes to India were proposed, hydraulic systems were invented, animals and plants were classified, and the functions of various internal organs were discussed. Important advances in number theory, geometry, and trigonometry were also made. As long as the libraries of Alexandria flourished, so did the fountain of discovery. But the museum and libraries were vulnerable to potent destructive forces, including fire, flood, and political or economic neglect, and their eventual demise in the fourth century A.D. coincided with a decline in the intellectual life and prosperity of the great city.

For discoveries and inventions to have an enduring impact on civilization they must be recorded and disseminated to those able to act upon them. Systems of communication influenced the adoption of Gregor Mendel's ideas. Mendel's findings were published in the obscure *Journal of the Brünn Society of Natural Science* and were widely disseminated only after other scientists acknowledged and confirmed his work 34 years later in more popular journals. Under slightly different networks of communications, his pea experiments could have either been forgotten entirely or have been immediately acted upon. Nowadays, an enormous quantity of innovative thoughts, experimental findings, and creative works are instantly captured and disseminated via computers. Electronic libraries of biological information are now integral to research in life sciences.

In 1979 a group of biologists and mathematicians met at Rockefeller University in New York and proposed that a database be established to hold DNA sequences.[1] A growing number of researchers were sequencing DNA and wished to more easily compare their sequences to those characterized by others. In 1981 the European Molecular Biology Laboratory (EMBL, funded by a collection of European nations) established the EMBL Data Library and in 1982, after many deliberations, the U.S. National Institute of Health (NIH) backed a proposal by Michael Waterman and Temple Smith of the Department of Energy's (DOE) Los Alamos National Laboratory in New Mexico. GenBank was established at Los Alamos on a five-year, $3.5 million budget.

GenBank is not a depository of genes, at least not in the physical sense. It is a repository of DNA sequences and information related to these sequences. Sequences in GenBank were culled from journal articles and entered into a computer, thereby allowing them to be stored, transmitted, analyzed, and displayed electronically. A new division of the National Library of Medicine (NLM), the National Center for Biotechnology Information (NCBI) was established to run the GenBank database. NLM already had over a decade of experience with MEDLINE, a database of articles from medical research journals.[2] Initially, GenBank was distributed to large computers at research institutions and updated periodically.

Starting in 1986, the National Science Foundation provided electronic links that connected computers at various academic departments among research universities. Thereafter, GenBank became
available through this forerunner of today's Internet. MEDLINE
was also put online and eventually the two databases were linked.

A handful of other databases also emerged in these early years,
including Swiss-Prot, a protein sequence database funded by the
Swiss government,[3] and an online version of Dr. Victor McKusick's
book *Mendelian Inheritance in Man* (OMIM). OMIM, which is
supported by the NLM and the Howard Hughes Medical Institute, describes all genes that are known to be linked to particular
human diseases and traits. These databases, as well as the instantly
popular e-mail services, helped medical researchers and other biologists establish an early and comparatively large Web presence,
which in turn helped kindle the explosive growth of the Internet.
For the research community the network provided instant access to
an ever-growing knowledge base. Rather than being cloistered in
voluminous libraries or confined within the craniums of a handful
of experts, details of the latest medical advances and gene discoveries were now literally at the fingertips of hundreds of thousands of
doctors, clinicians, and researchers. It is an example of governments
at their very best, providing resources that benefit a great many yet
cost only a minuscule amount. By making the databases extensive
and readily available, these government institutions eliminated the
need for redundant funds.

By 1985 GenBank contained over five million bases in close to
6000 sequence entries drawn from 4500 published articles.[4] However, the task of copying sequences from journal articles into a
computer was rather cumbersome, and as the practice of publishing novel sequences became increasingly popular GenBank faced
a crisis.[5] Its initial funds were quickly exhausted. Eventually a
much larger budget was approved and the data entry process was
changed. It became the researcher's responsibility to directly and
electronically submit newly found sequences to GenBank. Most
scientific journals supported this practice by making it a requirement for publication. Patented sequences and sequences that had
become public through international patent filing procedures

were also entered into the database. In addition, EMBL, the DNA DataBank of Japan, and GenBank cooperated in sharing DNA sequencing data.

Sequences alone are of limited value. The cells from which they were derived may have all they need to interpret or use the sequence, but we do not. So, along with a series of As, Gs, Cs, and Ts, researchers were urged to provide GenBank with additional information, such as the sequence's history (what species of animal it was derived from, etc.), key features of the sequence (the protein start codon, etc.), references to journal articles in which the sequence is discussed, and the names and affiliations of the discoverers. These annotations were linked to each sequence entry. A researcher could search the database for sequences that match a sequence of interest and receive a listing of those sequences along with their annotations. Several search programs were developed and bestowed upon eager researchers. The most popular was BLAST (Basic Local Alignment Search Tool), a blisteringly fast set of algorithms developed in 1990 by David Lipman, the first director of NCBI, and colleagues.[6] Database searches, a strictly computer-based activity, were performed thousands of times a day all over the world. Their impact on the course of research was profound. Suppose that you came upon a fragment of DNA that causes a particular cell type to divide without the normal constraints on growth (i.e., in a cancerlike manner). You can determine the sequence of the DNA at your laboratory bench and then query GenBank for sequences that match. Articles about any identical or similar sequences and the names of their authors and affiliations become immediately available. If you find an identical sequence, then you must consider how your findings fit in with what has already been reported. Do your findings lend support to previous ones, uncover a new role for the molecule, or contradict what others have found? These considerations may lead you to pursue additional experiments—or to abandon the project entirely. If you find a similar but not identical sequence, you may consider the nature of the similarities and differences and their relationship to the phenomena being studied. For gene hunters and other molecular biologists, GenBank, BLAST, and the computer network were godsends. From the mid-1980s, every gene discovered was queried against the GenBank database. The arrival of BLAST and GenBank helped bring forth innumerable discoveries and intensify biology's gene-centric focus.

In GenBank's early years, techniques such as DNA synthesis and polymerase chain reaction (PCR) were becoming part of almost every molecular biology laboratory. DNA synthesis machines enabled the rapid production of short (typically under 50 nucleotides long) single-stranded DNA molecules of any chosen sequence. With PCR billions of copies of any piece of DNA could be made, so long as it was relatively short (typically under 10,000 base pairs) and so long as the sequences of at least 20 or so nucleotides at each end were known. The chemical reactions that replicate DNA in cells could be made to occur millions of times in a single tiny tube, initiated by synthetically made strands of DNA. These techniques enabled scientists to readily apply sequence information itself, drawn from a computer database, to clone a particular gene, make sensitive assays for it, and conduct countless other manipulations. Thus, an electronic transmission could enable a scientist to rapidly obtain a physical copy of a gene discovered half a world away. The scientist at the receiving end must only have an animal or tissue source from which to extract DNA containing the desired sequence. But even this wasn't absolutely required. Although it is rather impractical, one could synthesize an entire gene, or even the sequence of an entire virus, from basic chemical building blocks, molecules that have never been part of a living organism. Currently, DNA without a cell is as forlorn as an embryo without a mother. So in no way can life be electronically transported. For better or worse, however, these relationships are in flux. Further insight into these issues will be provided in later chapters.

The sequence database itself proved to be a subject worthy of research. It was used, for example, to deduce sequences patterns associated with important functions. In one of the first scientific papers to reference the GenBank database, Kotoko Nakata, Minoru Kanehisa, and Charles DeLisi described a method for distinguishing the boundaries between coding and noncoding regions of DNA based on a statistical analysis of these regions in the database.[7] Many more such studies would follow. Anyone with an Internet connection may try such *in silico* research, as it requires no laboratory (*http://www.ncbi.nlm.nih.gov/*).

Blessed as it was by the good will and foresight of both administrators and computer scientists, GenBank was vulnerable to a type of neglect. Its most significant weakness lies in its system of quality control or lack thereof. Although it reduced costs, user entries led to redundancy, errors, and worst of all, misinformation. Researchers have been able to enter all sorts of fragments and variations of sequences already in GenBank. Annotations vary widely in terms of the vocabulary used to describe the features associated with the sequence. Different groups of researchers often arrive at the same gene and give it a different name. In addition, although some sequencers carefully verify their sequence by resequencing the same stretch of DNA several times, other may not do so. Sequence quality can vary. GenBank stipulates that an accuracy rate of 97 percent is required; up to 3 percent of the bases may be denoted as an *N*, an ambiguous base. For a researcher looking at his or her particular gene of interest, these data quality issues are not a major problem, because one can usually sort out most of the redundant or misleading information through careful examination or through replication of experiments, which in research is generally considered a necessary activity, anyway, rather than a waste of time. But for large-scale projects in which this data will be processed for use in other applications, the quality of GenBank data becomes a significant obstacle.

Despite its drawbacks, more and more researchers used GenBank. This is testimony not only to gene sequencing mania, but also to a property that databases share with libraries, businesses, and other sources of goods, information, or services. The chance of a user finding what he or she is looking for at a supermarket, mega-drugstore, huge bookstore, or major library is greater than at a smaller establishment. An early advantage in the marketplace can be used as leverage to propel further growth. As GenBank grew bigger (contained more sequences) and more comprehensive (covered a greater ratio of all known sequences) its usefulness climbed dramatically. This in turn attracted more users. The NCBI managed to both keep up with the volume of sequences and to slowly provide better quality controls and improved search capabilities. GenBank quickly became a convenient one-stop shop for DNA sequences and a springboard for a growing number of sequence analysis studies. By the end of 1990 there were 50 million base pairs in the GenBank DNA sequence database in over 40,000 entries from an enormous global network of scientists (see Fig. 5.1). It is hard to imagine anyone competing with this. Nonetheless, this is exactly what happened.

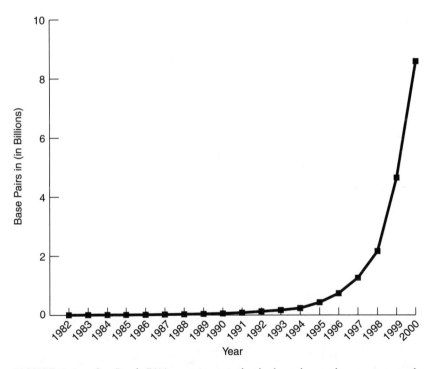

FIGURE 5.1 GenBank DNA sequence submissions have risen astronomically, as is illustrated by this chart. Many entries in this massive database are redundant (particularly the EST sequences), and while the majority are from human beings, over 50,000 other species are represented. *(Data from NCBI, www.ncbi.nim.nih.gov/Genbank/genbankstats.html.)*

Before considering databases that would rival GenBank, let's take a closer look at who is sequencing DNA, the types of sequences they are producing, the ownership of these sequences, and their uses. In the classic 1963 film *It's a Mad, Mad, Mad, Mad World*, a set of wacky contestants with crude maps races toward a distant buried treasure. With the scientific and commercial value of genes well established, a stage was set for an equally wacky contest. The goal is more noble and historic, but there has been plenty of madcap action, interesting characters, and shifting alliances. The heated contest for human genes has brought forth a variety of intertwined scientific, legal, and business strategies. It has also raised many hot issues, such as the proper roles of governments and businesses in medical research and drug development, the question of ownership of biological material and information, the nature of humankind's genetic heritage, the potential for doing biological harm, and the prospects for improving human health and other desirable things. These are issues that affect all of us and ought to be widely examined and actively debated. An account of what has transpired so far and a description of what has been learned may help provide a better basis for such debates. Tally ho!

6

Getting the Genes

When the height of Mount Everest was measured in the middle of the nineteenth century, it came to precisely 29,000 feet, not a foot more and not a foot less. The geographers who determined this felt that this result hid the great lengths they had gone to get an accurate measure. To make sure that their result would appear precise, they falsely restated the height as 29,002 feet. Ironically, later and reportedly more accurate measures put it at 29,028 feet. The number of human genes will never be measured with such precision. One hundred thousand is the most-often-cited estimate of the total number of different human genes. It is a nice round number. Beyond this, it is not very clear why this number has been so often used.[1] So why won't we ever really know precisely how many genes there are? First, let's consider what constitutes a gene.

Once the structure of DNA was unraveled and the genetic code was deciphered, it became possible to determine a physical definition of a "gene," going beyond its original concept of a heritable unit of biological information. Genes are particular stretches of DNA which, when perturbed, result in observable consequences, such as a change in protein abundance or composition, a trait, or a disease.[2] In most cases genes encode proteins with RNA molecules acting as intermediaries. However, there are a few dozen genes that code for RNAs which do not yield protein, but perform structural or catalytic functions themselves. The stretch of DNA that constitutes a gene may also include sequences that directly influence RNA production (*transcription*) or protein production (*translation*) although they do not themselves encode protein. In humans and other multicellular organisms, protein-encoding sequences most often exists in fragments (*exons*) interspersed along longer stretches of DNA. The longer stretch of DNA is made into RNA and the intervening sequences (*introns*) are subsequently removed (*spliced out*) (see Fig. 6.1). In addition, exons may be combined in alternative ways, creating distinct but related mRNAs known as *splice variants*. The resulting proteins may differ only slightly but nevertheless have entirely distinct roles. Even so, they are often described as originating from the same gene. Strictly speaking, a stretch of DNA that performs a structural role may also be considered a gene. For example, there are several hundred base pairs of DNA that constitute the centromere of the chromosome, acting as a handle for the molecular apparatus that pulls replicated DNA strands into different portions of the cell in advance of cell division. However, most gene analyses disregard these nontranscribed stretches of DNA. In general, each gene results in a distinct RNA with a distinct function.

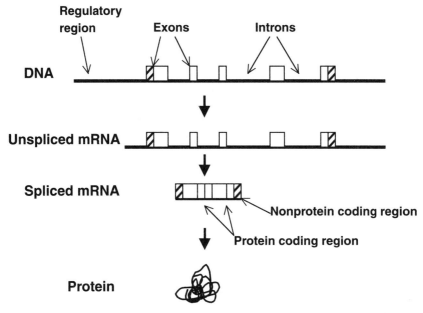

FIGURE 6.1 Gene expression. Genes are encoded in DNA sequences (shown as a line). RNA is transcribed from DNA, and intervening sequences (introns) are spliced out. Mature mRNA consists of the remaining sequences (the exons, shown as boxes). The mRNA sequence is translated into an amino acid sequence (which makes up protein molecules). A portion of the mRNA sequence at each end (striped boxes) does not encode amino acids.

The entire set of genes and the sequences that surround them is known as the *genome*. The genome can be considered as the information content of an organism, the full set of inherited instructions that dictate the form and manner of life from conception onward. However, this is not entirely accurate, for it is known that certain factors outside of the DNA sequence may be carried into new life via the egg or sperm. These *epigenetic* factors may make significant contributions to genetic diseases and traits.[3] It so happens that genes influence these epigenetic factors too, and so most biologists seem quite content with life's central stage being occupied by the full set of genes. The genome is considered as being an organism's "operating system."

The genes of most bacteria lie on a single circular DNA molecule, which thereby constitutes the bacterial genome. Genes in other organisms most often reside along several linear strands of DNA. At the time of cell division the strands condense around various proteins to form chromosomes, structures that facilitate the equal distribution of replicated DNA among daughter cells. The DNA component of chromosomes plus multiple copies of relatively minuscule circles of DNA in mitochondria or chloroplasts (energy-producing components of cells that are believed to be derived from bacteria captured billions of years ago) compose the genome of sunflowers, people, rats, ants, coyotes, and so forth.

With only a few exceptions, the complete genome is contained in all cells of a multicellular organism; thus a neuron, a muscle cell, a white blood cell, a kidney cell, and a skin cell will have nearly the same genome. This is why an udder cell of a mature sheep can be used to generate an entirely new lamb, as was demonstrated by the creation of Dolly, the famous cloned sheep of 1997. Strictly speaking, however, since the fidelity of DNA replication is not perfect, the genomes of different cells within a single organism (and even of identical twins) are not precisely the same. Genome variation due to imprecise DNA replication is noteworthy because this is believed to be a significant factor in the development of cancer, in the development of genetic diversity, and in the evolution of species.

The expression of distinct sets of genes underlies the differences between cell types. A rough fix on the number of distinct human genes expressed in particular tissues was found in the late

1970s by way of hybridization kinetics studies. The output of a genome, the genes it is expressing at a particular time, are transcribed into mRNA. A sample of mRNA was collected and converted into double-stranded DNA, known as complementary DNA (cDNA). Each cDNA can be readily purified and manipulated, because it can be spliced onto a circular DNA molecule that can be propagated in bacteria. In other words, the cDNAs can be readily cloned. A collection of cDNAs from one group of cells is known as a cDNA library.

The strands of DNA from a cDNA library were made to come apart, and the rate at which they rejoined (*hybridized*) was measured. Amid this mix of DNA strands, those strands derived from highly abundant mRNAs will find a match rapidly, while strands derived from rare mRNAs will take much longer. It is as if you were lost at Disneyland. If you came with just one other person you would have a far more difficult time finding that person than you would have finding any one member of a large and dispersed group of friends. The time required for all DNA strands to find a partner depends upon the number of different types of mRNA. The greater the diversity of mRNA types, the longer it takes for all the various DNA strands to pair up. This diversity is also known as the *complexity* of the transcribed genes.

Using measurements of hybridization times, it was determined that liver cells have about 10,000 different types of mRNA representing the expression of 10,000 distinct genes.[4] Very different cells, those of a cancerous blood cell, were also found to express 10,000 distinct genes, with between 75 and 85 percent of the sequences identical to those in the liver cells. Brain tissue was found to express as many as 30,000 distinct genes, and it was suggested that the complex cognitive functions that the brain cells perform require a greater number of active genes then do the functions of other organs.[5] From these and other hybridization kinetic studies, it was estimated that the total human genome consists of 90,000 genes.[6]

Another approach to determining the total number of human genes relies on some of the genome's physical specifications. The human genome is contained in DNA weighing about 2×10^{12} Dalton, the mass of the DNA in a single cell.[7] It consists of two copies of about three billion base pairs of sequence, each arranged on 23 chromosomes. If the DNA from one cell were untwisted and strung out, it would measure about 6 feet in length. But the protein coding sequences of a typical gene only contain about 3000 base pairs. Thus, three billion base pairs are sufficient to encode a million distinct proteins.

Human genes are not packed so densely, however. Sequencing studies involving just a very minute portion of the total genomic DNA have revealed that within a single gene the introns, the non-protein coding regions, most often occupy tens of thousands or more base pairs, and the distance between genes is often much greater than this. A very small fraction of the sequence between protein coding regions has been shown to regulate gene expression (i.e., to guide the agents needed to transcribe RNA from DNA, splice RNA, and synthesize protein). The remaining DNA sequence has no apparent function; it appeared to be "silent." An average of one gene was found for every 23,000 to 70,000 base pairs, depending upon which region of the genome is sequenced.[8] To arrive at a total of 100,000 genes, one need only to assume that the overall average gene density was one per 30,000 base pairs. These 100,000 genes would occupy about 3 percent of the genome.

What about the approximately 97 percent of the genome that does not encode protein? Passions arise in describing this vast region of DNA. It is *our* genome we are talking about here, the holder of our biologic uniqueness, the family jewels we pass on to our children. Could 97 percent of this treasure be useless? People have, in fact, been found possessing millions of base pairs less DNA than others, yet showing no ill affects.

The "surplus" DNA has been vilified by some as garbage, but has been deified by others as the holder of untold mysteries. The struggle over this DNA's "reason for being" largely began with a 1972 article describing this DNA in several species. The article was entitled "So Much 'Junk' DNA In Our Genome."[9] For an example of junk one need only to have looked at the dull, ubiquitous, short stretches of DNA known as *Alu* sequences.[10] Alu sequences are approximately 300 base pairs in length and occur almost a million times in the human genome with only a small degree of variation. There are several other types of repet-itive sequences, some as monotonous as repetitions of the nu-cleotide pair CA. In total, about one third of the human genome is believed to be highly repetitive sequences, while almost two-thirds is thought to be nonrepetitive sequences, which, by a variety of tests, have nevertheless been shown not to encode pro-tein.

Many scientists were reluctant to dismiss any DNA as use-less. Indeed, the notion that a billion of years of evolution would yield "junk" was offensive to some people. Numerous possible functions for these sequences were proposed. For example, by providing abundant sites where DNA strands can safely be cut and rejoined (recombined), these sequences may help create beneficial combinations of genes. Or, some of this DNA could have a structural role, somehow providing a more stable chro-mosome framework. Or, some of these sequences may act as spacers between regulatory sequences and protein coding se-quences, analogous to the spaces between words in these sen-tences. However, experimental evidence supporting these theories dribbled in only slowly and could not conclusively ex-plain most of the noncoding DNA. Matching a function with all these sequences has been as difficult as finding the missing dark matter of the universe, that postulated stuff upon which the fate of the universe may depend. No one knew just where to look or what tests to apply. Perhaps the effects of these sequences are so subtle that they only become apparent over the course of many generations, or perhaps they performed important roles only in the distant past. The latter possibility was proposed by Susumo Ohno, a scientist at the City of Hope Medical Center in Duarte, California, and the author of the 1972 article that introduced the notion of "junk" DNA. Ohno had suggested that these se-quences may be the remnants of former genes: "The earth is filled with fossil remains of extinct species; is it a wonder that our genome too is filled with the remains of extinct genes?"[11] It has also been suggested that some of these sequences may help to provide for future generations. For example, they may provide the raw material for new genes.

All of the functions just stated suggest that, at least at some point, these sequences provide a benefit to the organism by enhancing its ability to survive and reproduce. But what if DNA sequences did not serve the organism in which they dwell? What if DNA sequences were to follow their own agendas, their own survival needs? Evolutionist Richard Dawkins explored this idea in his 1976 book, *The Selfish Gene*.[12] The notion of parasitic DNA was inspired by the behavior of viruses, which can integrate their DNA into the genome of their host and manipulate molecular processes in the host to support their own survival. Genomic sequences can be just as insidious. Many repetitive sequences are flanked by viral-like sequences that support both the independent replication of the sequences and the integration of copies elsewhere in the genome. They are known as *retrotransposons*.[13] Selfish DNA may persist in the human genome by virtue of one or more of the following: the sequences may be harmless or nearly harmless to their host and may simply piggyback from parent to child along with other, useful sequences; they may be readily generated like weeds and thus appear to be ever-present; or they may mutate often or have other defensive postures that prevent their elimination.

Ohno later provided a more comprehensive explanation for different types of genomic DNA. He proposed that DNA sequences go through their own life cycle: they are born "selfish" (through events previously described), turned into useful genes, and then after new genes or new environments render them useless they slowly degrade into "junk" DNA. They may be readily eliminated during the beginning and end phases, but the cycle could last millions of years.[14] Genes are thus somewhat like wealth. They may come about by self-interest and opportunity, persist when they are used well (as in sound investments), and then eventually become diluted out or squandered.

Decades later the debate over the function of nonprotein coding DNA continues. Fortunately, new research findings are shedding more light on the issue. For example, in 1998 a team of researchers found convincing evidence that at least some of the Alu sequences function in cellular responses to certain stresses.[15]

Alu RNA is made in response to viral infection, and although this RNA is not made into protein it affects cells' overall production of proteins through its interactions with the protein translation apparatus. This in turn helps cells withstand infection. Some portion of the repetitive DNA therefore functions in a way that serves the organism. Clearly, there is much more to be learned about the role of nonprotein coding DNA.

So much for now about the "biological" value of genomic DNA sequences. What about their value outside of the genome, their ability to impact human lives? Despite the misgivings, Alu sequences and their kin have been the subject of well over a thousand scientific papers and have yielded clues to questions of how mutations arise, how mutations are spread through a population, the nature of evolutionary change, and the history of life. Genomic DNA, of course, also harbors the genes that have proven value to human health care, and by the mid-1980s interest in the human genome's three billion base pairs of DNA was sufficiently high for large-scale sequencing plans to germinate.

Humankind has endeavored and succeeded at goals far more difficult than that of sequencing the human genome. Enormous pyramids have been erected, oceans have been crossed, people have been sent to the Moon and back, atomic energy has been harnessed, the deadly smallpox virus has been practically wiped off the face of the earth, would-be world dictators have been snuffed out, and wars in Vietnam, Bosnia, and Palestine, for example, have been stifled. These achievements provide historic milestones, data points in the study of human history. They have also, at times, captivated the minds and enlivened the spirits of contemporaries, as the U.S. space missions did for Americans in the 1960s and 1970s.

U.S. national priorities appear to have shifted in the 1980s and the 1990s. Star Wars (the purported plan to build a space-based strategic missile defense system) and the superconducting supercollider (an enormous atom-smashing facility that had been partially constructed in Texas) were abandoned, and a plan to map and sequence the human genome gained support. National defense and nuclear physics objectives were out of favor; human health needs and molecular biology were in.

The initiative to sequence the human genome was remarkable for biological research. First, it represented a bottom-up approach. In genetics the traditional approach has been to start with a heritable disease or trait and work down to the genes involved. Reverse genetics, where one starts with a gene and then determines what happens when it is perturbed in various ways, was new in the 1980s. Second, the initiative was in the category of so-called "Big Science." "Big Science" was any research project that required sharing of a single costly resource, such as a linear collider or a satellite. "Big Science" had previously been the exclusive domain of the physical sciences, where monstrously complex and costly instruments were deemed necessary to probe the mysteries of the atom, the earth, or of outer space. In biology small, independent, and self-sufficient laboratories were the norm. Most biologists had probably never contemplated a billion-dollar project. Nevertheless interest in human gene sequences and anticipation of their future importance grew so great that by the mid-1980s an initiative to sequence the entire human genome was ready to take root.

At two separate meetings—one in 1985 and the other in 1986—Robert Sinsheimer, the chancellor of the University of California at Santa Cruz, and Charles DeLisi, the director of the Office of Health and Environment for the Department of Energy, rallied influential scientists to the cause. Meanwhile, Renata Dulbecco, a world-renowned cancer researcher and president of the Salk Institute in San Diego, California, declared the importance of a human genome project in an editorial in *Science* magazine. The first concerted effort to sequence the human genome emerged as a consequence of these exchanges.

There was much spirited debate over the nature of the project, which concerned scientific, political, and economical issues.* Among the scientific concerns expressed were the relatively low yield of genes in genomic DNA, the inadequacies of sequencing technology, the difficulty in making any sense out of three billion

---

*An excellent chronicle of the events appears in *The Gene Wars*, a 1994 book by Robert Cook-Deegan, a physician who had directed a study of genome research for the U.S. Congress' Office of Technology Assessment.[16]

base pairs of sequence, and the need for gene maps. In the journal *The Scientist*, Robert Weinberg of the Whitehead Institute noted that "a gene appears as a small archipelago of information islands scattered amidst a vast sea of drivel," and by sequencing the genome researchers would be "wading through a sea of drivel."[17]

Several people, including Sydney Brenner, a developmental biologist at the University of Cambridge, pointed out that the most direct route to human gene sequences is through mRNA. The cell has a mechanism that bypasses the large portion of the genome that does not encode protein. Cells produce mRNA, which is dense with protein-coding sequence and which can readily be made into cDNA and sequenced. It was a noteworthy suggestion.

At a 1986 meeting at the U.S. National Institute of Health, Leroy Hood, a professor at the California Institute of Technology and inventor of automated DNA sequencing technologies, asserted that massive sequencing was premature "and that the focus instead should be in improving the technologies."[18] At a 1987 roundtable forum Francisco Ayala, a mathematician and professor of genetics at the University of California at Davis, stated that "it would not be feasible at present to attempt to decipher and understand a genome as large as that of the human being. Mere size puts it beyond current possibilities. . . . Testing the paradigm of molecular biology is a very long-term enterprise, but we could start with *E. coli* [a bacteria with a genome about one one-thousandth the size of that of *Homo sapiens*] in perhaps less than a decade from now."[19]

Knowing the chromosomal location of a large set of sequences that vary throughout the human population would greatly help in pinpointing more disease- and trait-causing genes. Such chromosomal maps are also necessary for identifying, producing, and organizing the set of over three million DNA fragments that would yield the sequence of the entire genome. Walter Bodmer, a population geneticist and director of the Imperial Cancer Research Fund of London, urged that mapping the chromosome be given a high priority.

Among the political and economic concerns were the cost of the project, the impact on other scientific projects, and the impact on society. David Botstein, a professor at Massachusetts Institute of Technology, cautioned against changing "the structure of science in such a way as to indenture all of us, especially the young people, to this enormous thing like the Space Shuttle, instead of what you feel like doing."[20]

All these concerns would eventually be redressed, at least to some extent. The initiative had enough charisma for momentum to build further. Government and/or private foundations supported genome projects in the United Kingdom, Japan, France, the Soviet Union, and Italy. In the United States large-scale human genome sequencing plans advanced at two separate government agencies and at one nascent company. DeLisi and colleague David Smith saw human genome sequencing as a means to fulfill the DOE's mission of understanding the biological effects of radiation (a carryover from the post-Hiroshima and Nagasaki bomb blast era) and of more fully utilizing the DOE-administered labs. In fall of 1987 the DOE established three genome centers at U.S. national labs. In that same year James Watson, the director of Cold Spring Harbor Laboratory, David Baltimore, a scientist at the Whitehead Institute, and NIH director James Wyngaarden asked for and received appropriations from Congress for NIH-directed genome research. The DOE and NIH efforts were soon enjoined.

Meanwhile, Walter Gilbert, a Harvard professor and a very active early participant in the human genome sequencing brouhaha, made his own attempt to initiate a human genome project. Gilbert, who had recently cofounded the biotechnology company Biogen, figured that academic and national laboratories were not the most appropriate place for large-scale sequencing. Academic scientists would be better off focusing their efforts on science rather than grunt work, he reasoned.[21] In the spring of 1987 Gilbert announced plans to form Genome Corporation, which would map and sequence the human genome through the use of industrial-scale processes, automation, and computers. Genome Corporation was to sell access to a resulting sequence and mapping database, do contract sequencing, and sell gene clones, the physical matter, to interested parties.[22] However, prospective investors were leery of competing with what could be a high-budget government-backed effort and were especially cautious following the stock market minicrash of 1987. Genome Corporation never took flight, and Gilbert returned his attention to the publicly funded effort, where he extolled the same vision of factory-like sequencing operations and computer-based analysis.[23]

In Europe, Renata Dulbecco inspired the Italian government to back a human genome project, while Sydney Brenner instigated a genome project in the United Kingdom with funding from the government's Medical Research Council (MRC), the Imperial Cancer Research Fund (a private foundation), and from his own accounts. In France a foundation for muscular dystrophy helped launch a program in human genome research, and authorities in Japan and the Soviet Union also supported significant human genome research. Brenner and Victor McKusick helped start an organization that would coordinate all the international sequencing and mapping efforts. The Human Genome Organization (HUGO) was created in 1988 with funding from private foundations, the Howard Hughes Medical Institute, and the Imperial Cancer Research Fund.

Gilbert had figured that it would cost about $3 billion to fully sequence the genome, a dollar for each nucleotide. A committee commissioned by the U.S. National Academy of Sciences and the National Research Council agreed and suggested outlays of $200 million a year for 15 years. Much of the projected expense would be in preparing an enormous set of overlapping DNA fragments for sequencing, rather than the sequencing itself. The Human Genome Project (HGP) was formally launched in October of 1990 under direction of the DOE, but with the participation of the NIH. The early emphasis would be on developing better sequencing technology and creating a denser map of sequence variations. A program was set up to explore the ethical, legal, and social implications of human genome research; to select which model organisms would be fully sequenced; and to deal with computational issues, such as how to store, transmit, display, and analyze the sequences. In the later portion of the 15-year plan, the human genome itself would finally be fully sequenced.

The U.S. commitment to biological and medical research was greater than that of any other nation, and the United States would maintain its leadership in the field of human genome research. In 1990 about $90 million was earmarked for genome research in the United States, a figure that would rise every year thereafter. By the end of 1990 everything seemed set. The "Holy Grail," Gilbert's term for the full complement of human genes, seemed to be in the bag. Various U.S. agencies, universities, and institutes were openly and productively interacting under the HGP umbrella. An international cooperative effort was in place that could put the United Nations to shame. Funds were available, great talent was in place, and countless scientists, educators, and members of the general public, with their hearts and minds, were onboard. Most of the objections to the early plans had been addressed. Remaining critics were left on the dock as the massive ship left port.

The Human Genome Project would yield an impressive list of accomplishments, but during its first 10 years it would not even come close to providing the final word in genome research. As will be described in coming chapters, innovative strategies and new technologies largely developed and applied outside the realm of the HGP would reveal the sequence and activity of more human genes than would the HGP and would open up revolutionary new approaches to scientific discovery and disease intervention. These projects had their intellectual foundation in the same flurry of discussions and debates that gave rise to the HGP. Thus, the climax of the HGP project may have occurred even before its huge infrastructure and budget were in place, for it was at this time that the seeds of a revolution in the life sciences were sown. Perhaps only a handful of highly attuned contemporaries may have picked up some sage thoughts that now appear as clear as a bell, such as those expressed by Akiyoshi Wada, a biophysicist at the University of Tokyo. In 1986 Wada wrote that automation in molecular biology "could well turn out to be the equivalent of the Industrial Revolution in biological and biochemical laboratories."[24] This prediction was right on the mark, as will be shown in later chapters. Wada embarked on a DNA sequencing automation project for Japan's Science and Technology Agency way back in 1981. He helped create a collaboration between government and industry in Japan and helped incite and inspire the U.S.-led HGP.

Walter Gilbert should also be granted sage status. The grand scale of his visions and his business aspirations may have led many to dismiss Gilbert as a hypester, but he clearly foresaw and helped lay the groundwork for a new framework of thinking. "To use this flood of knowledge, which will pour across the computer networks of the world, biologists not only must become computer-literate, but also change their approach to the problem of understanding life," he wrote in a widely-read article entitled "Towards a Paradigm Shift In Biology."[25]

By 1991 the debate and hype that the HGP had generated created a climate that begged for action rather than words. Let's now return to the subject of actually getting all the human genes, however many there may be.

7

Prospecting for Genes with ESTs

While the Human Genome Project set a course for obtaining all human genes by way of genomic DNA sequencing, several other teams set off on another tack. Craig Venter, who was with the NIH at the time but not supported by HGP grants, piloted an effort to obtain novel genes via mRNAs. Venter had once attempted to find the gene responsible for Huntington's disease on a 60,000 base-pair stretch of chromosome. During this unsuccessful search, Venter learned firsthand the difficulties in finding and piecing together protein encoding sequences by way of genomic DNA. Then, in a landmark study published in *Science* magazine in June of 1991, Venter and colleagues described a pilot program in which they partially sequenced 609 randomly chosen cDNAs derived from three samples of human brain tissue.[1] The sequences were long enough to provide a means to identify the cloned gene from which they were derived (a sequencing shortcut that was first described in two independent reports published in 1983) and were now termed *expressed sequence tags* (ESTs). (A typical mRNA is between 1000 and 10,000 nucleotides long; in the *Science* report the ESTs were between 300 and 350 base pairs, while later ones averaged only 150 base pairs). These rapidly produced tags were tentatively identified by a search of GenBank using the BLAST algorithm. The ESTs fell into four main categories: (1) those that were identical to a portion of a known gene, (2) those with sequence similarity to a known gene, (3) those which did not match anything in the database, and (4) those that could be deemed useless because they were either devoid of meaningful sequence or matched sequences of contaminating organisms. Venter and his colleagues found that 38 percent of the ESTs did not match any known sequence and thus were likely to represent novel genes. Only a small fraction of ESTs overlapped or otherwise matched the same gene. The researchers also demonstrated that they could increase the discovery rate by using cDNA libraries from which abundantly expressed transcripts were removed. By separating the DNA strands and eliminating those that rehybridize quickly, the newly reconstituted cDNA library yields an even greater percentage of novel genes. The procedure was rapid and scaleable. Thus, Venter and associates appeared to have proved an assertion made during the genome project debates. The quickest route to human genes was through mRNA. As of yet the effort to identify and characterize all human genes had generated many words, but few deeds. Now a stage was set for a much larger gene discovery effort. The initial set of ESTs was submitted to GenBank. What next? Scale it up and get all the genes.

It appeared so straightforward. With a somewhat more laborious effort the full gene sequence and even regulatory sequences could be derived from any EST. Rapid, high-throughput partial cDNA sequencing could be a springboard for getting every gene, mapping them, and determining their functions. No doubt, there were logistical hurdles to overcome, technological challenges that could be met with a multitude of innovative schemes. However, the events that followed Venter's proposal pointed up an entirely different issue, namely the glaring lack of central authority over the means, manner, and direction of research in the life sciences. A jostling crowd of different political, legal, business, personal, and other interests would contend in the madcap race to discover human genes. No one entity, no nation, no council of scientists, no medical foundation, no political group, no grass roots organization, no law, and no corporation could fully control or direct the pursuit of human genes. However, this seemingly chaotic state of affairs did not hinder the advancement of science or the development of new medicines. On the contrary, a lack of overriding authority probably helped enable experimentation to flourish and new discoveries to spring forth. Not only would virtually all human genes eventually be revealed; the groundwork for a profound new understanding of life would be laid as well.

The National Institute of Health filed a patent application on Venter's "invention," 337 unique members of the set of ESTs described in the 1991 *Science* paper, and a firestorm of protest ensued (see Appendix for FAQs about biotechnology patents). A second patent application with more than 2000 ESTs was filed in 1992. Respected leaders, such as then NIH Genome Project director James Watson, decried the "sheer lunacy" of it and expressed nothing less than "horror" over the very idea of patented gene fragments.[2] Critics contended that allowing gene fragments to be patented would make researchers more secretive and stop them from sharing data, thereby slowing the race to obtain the full sequence of all genes. They imagined that by virtue of an EST patent the patent owners could "lock-up" a gene and remove the incentive or the opportunity for others to further characterize it or develop it into a useful product. "No one benefits from this, not science, not the biotech industry, not American competitiveness," asserted David Botstein, who at the time was both the chairman of Stanford University's Department of Genetics and a vice president at Genentech Inc. The faith that many scientists and science administrators had in the U.S. patent system was below the limits of detection. David Galas, who headed the genome effort at the Department of Energy, contended that "there is no coherent government policy, and we need one quick since the sequence is just pouring out. . . . It would be a mistake to leave this one to the lawyers."

The criticism over these patent applications quickly carried over into broader sectors of society. EST patents were soon associated with the ownership of genes, the ownership of living creatures, human slavery, eugenics, and one of the most widely recognized of all modern day horrors, the Nazi-instigated Holocaust.[3] Acts of Congress and international treaties were proposed to put a stop to these patents. Meanwhile, on the receiving end of the patent application, the U.S. Patent and Trade Office (USPTO) had yet to consider the patent claims of Venter's discovery.

The USPTO's mission as outlined in the U.S. Constitution is "to promote the progress of science and the useful arts, by securing for limited times to authors and inventors the exclusive right to their respective writings and discovers."[4] U.S. patent laws and procedures have evolved over two centuries to promote the creation and deployment of useful innovations. Inventions are awarded the status of property, but with limits that are designed to prevent further innovation from being compromised. Reid Adler, the director of NIH's Office of Technology Transfer, countered critics by contending that patenting ESTs would offer the protection needed to persuade drug makers to license the inventions and develop new medicines. He said that if the NIH put the sequences in the public domain and did not patent them, then they would be rendered unpatentable and drug makers would have less of an incentive to work with them.

NIH officials were also influenced by the spirit of the Bayh-Dole and Federal Technology Transfer Acts of 1980 and 1986, respectively. These congressional acts were meant to ensure that federally funded laboratory scientists and engineers supported technology transfer and commercialization through patent protection and licensing. Taxpayers would benefit from such patents in two ways. They would lead to commercial products that were previously unavailable, and licensing revenue generated from such patents would help fund further research.

Neither the U.S. Constitution nor 200 or so years worth of patent rulings had foreseen the ability to copy and use heritable material extracted from human tissues, a material that happens to be intimately connected to the design and function of our very selves. The physical nature of the gene had been known for only a minuscule amount of time and the ability to manipulate genes has been around for even less time. Legal proceedings, which often take 10 or more years to manifest themselves, had not dealt with such material until the 1970s.

By the 1980s, legal trends were being established that raised considerable concern among many segments of society, including many scientists. In 1976 a patent was awarded for the use of the recombinant DNA technique, a process by which genes of any species can be introduced into (*recombined*) and propagated in microorganisms. Then, in 1980, the U.S. Supreme Court overturned a lower court ruling and allowed a patent for a genetically manipulated oil-consuming bacterium. By the early 1990s, the USPTO was awarding patents for many genetically altered forms of life and for many types of material derived from living things, including newly discovered proteins and genes.

Hundreds of years ago, native Americans are said to have been baffled by the concept of land ownership. In today's crowded world, many folks still wonder how it is that a piece of mother earth can be owned. Nevertheless, it is now nearly impossible to find a single patch of land that has not been claimed, and not just for a 20-year period but for eternity. Farmers and others may mercilessly exploit their land for their own benefit. Many people believe that this actually helps avert tragedy, as in "the tragedy of the commons," the idea that damaging overuse and misuse occurs when no one in particular owns the land. Certain forms of life have also been kept as property. Breeds of cattle, sheep, and cats, strains of beer-producing yeast, and highly desired plants have been withheld from others so as to produce greater profits for their owners.

Despite these precedents, the advent of powerful new gene manipulation technologies coupled with the enforcing support of the legal system introduced a whole new era in the private ownership of living things. The patenting of life forms has made many people uneasy—or worse. Many feel that it presumes that the patent holder created life, usurping the role of God or some other higher power. The patenting of the molecular components of life has been similarly unsettling. How could genes or proteins that nearly all of us produce naturally be credited to an individual or group of people? It appears as if patent offices are granting "Creator" status to human beings for their gene discoveries. This is highly offensive to a large number of people.

Nonetheless, many of these same people also recognize the healing potential of these so-called inventions. The fact that you and I may make insulin does not help a diabetic person one bit. A bacteria engineered to produce human insulin does. The USPTO acted in accordance with its mission by assuring an incentive for people to make new gene-based medicines, such as bacterially produced human insulin. Furthermore, in no way did the USPTO restrict anyone's internal production or use of their own genes. Despite these facts, the idea of patenting the material of life continues to disturb many people.

For an invention or discovery to be patented the USPTO must be given sufficient reason to believe that it can be made and that it will be useful. This meant that initially researchers had to physically make a gene product, such as purified human insulin, and demonstrate its usefulness. During the 1980s successful full-scale clinical trials were often necessary to convince USPTO examiners that a particular gene discovery was patentworthy. But as gene sequencing efforts advanced, pressure grew for a change in policy. Isolated genes and gene products had utility as research tools. Incentives for the development and distribution of such tools could hasten the drug development process. After formal hearings in 1994, the USPTO issued new guidelines for their examiners. No longer were they to act like FDA regulators assessing the efficacy of new drugs; now inventors had to provide only a credible assertion of utility.

The hurdles for gene patenters were also eased by ongoing advances in technology. Throughout much of the 1980s, the act of producing an active and purified protein from a given gene sequence was an art form. By the 1990s, what was once a complex and unpredictable process had become nearly routine for skilled researchers. And the Patent Office requires nothing more than a written description for such routine procedures. Furthermore, the ability to predict protein function from sequence information had advanced. The outcome of a BLAST search, for example, could be used to predict that a novel sequence is likely to produce a protein with cell-growth-promoting properties or a particular enzymatic activity. Thus, the gene sequence itself, Mendel's powerful little unit of information, came to be the essence of the gene patent.

The National Institute of Health expressed sequence tag application sought to bring the basis for drug discovery claims one notch lower. It explained that an EST could be used to obtain the full protein-coding region of a gene by processes that were rapidly becoming routine. However, in breaching the gene level, it appears as if a line in the sand had been crossed—as if truly sacred territory had been disturbed. The slippery slope toward easy gene patenting suddenly became sticky. The PTO promptly rejected the EST patent claims. Actually, it initially rejects most of patent claims it receives, but in this case the NIH chose not to pursue the issue further. Many other EST applications were filed, but the combination of a slow-moving Patent Office and strategic stalling on behalf of patent applicants left the issue of partial gene sequence patents unresolved and smoldering for years thereafter. One key issue was the lack of a written description. Inventions must be fully described in the patent application, and a full description of a gene sequence is not provided by an EST. When patent applications on EST sequences were finally allowed in 1998, protection was limited to the partial gene sequence.

For years, not only was there no authoritative judgment either allowing or disallowing gene fragment patents, but there was also no congressional action or international treaty; and yet, as will be shown in coming chapters, the pursuit of human genes and gene-derived medicines did not appear to be hindered at all. Forces far stronger than that of patent law may be looming over molecular biology research. The desire for better medicines is extremely strong, and the most compelling means of pursuit of better medicines is through molecular biology research. This may be the dog that wags the patent tail (and often the legislative tail, the presidential tail, clearly the corporate tail, and the tails of many, though certainly not all, mortals with concerns over sickness, pain, and death). There are even at least two Congressional Acts that specifically limit the reach of patent law in medically important matters. A provision in the 1980 Bayh-Dole Act, the so-called "march-in-rights," enables the government to seize control over certain patents if health care is thought to be impeded by either the actions or inaction of those holding rights to the patents. And the 1984 Waxman-Hatch Act enables patent infringement litigation to be halted while drug products are in clinical trials. As of yet, the U.S. government has not felt compelled to use the march-in-rights of Bayh-Dole, but Waxman-Hatch has been invoked several times. The message is clear: business and patent issues will not stand in the way of new medicine.

In the end, patent law seemed to fulfill its Constitutional mission. Gene discoverers, or rather the institutions for which they worked, were rewarded with patent protection on sequences that encompassed entire protein coding regions, so-called "full-length genes." Competition for gene discovery was driven, in part, by a desire to patent and/or to prevent others from patenting. Even the ongoing uncertainty over EST patents, seemed to have spurred, rather than hindered, progress in uncovering gene sequences, determining their functions, and creating useful products. Patent laws were interpreted and applied in ways that promoted gene discovery and innovation. (See the appendix for more information on how the patent process works.)

Patents aside, what about Venter's gene discovery scheme? His one-lab effort was churning out up to 150 ESTs a day in 1992, and a fairly large percentage of them presented information on previously unknown genes. Initially, neither the decision makers for the Human Genome Project nor those with other government agencies embraced the EST plan. Perhaps, the public outcry against the patent application spoiled the prospects for a large-scale EST program. Watson likened Venter's work to something that "monkeys could do," and the influential Stanford biochemist Paul Berg was quoted as saying that the scheme "makes a mockery of what most people feel is the right way to do the Genome Project."[5]

The idea of accessing genes via cDNA had been brought up in early HGP discussions and rejected. Far from spurring a revision, the current EST controversy provided an opportunity to restate and further clarify the HGP's original genomic mapping and sequencing goals. In April of 1993 *Nature* magazine offered the following editorial opinion:

> . . . the two approaches to the structure of the human genome are neither in conflict nor complementary but different. The classical human genome project, figuratively (only) that of starting from one end and working through to the other 3 billion base-pairs later, will yield information the cDNA technique cannot. To abandon the first because of the second will yield immediately useful information would be like trying to teach a person a foreign language by instructing him only on the meanings of the nouns. The difficulty, at least until the dust has settled, is that the cDNA sequences, which would be of great value to the

larger project, are likely to be generated most rapidly under the umbrellas of companies and may not be published until patent rights are granted, whenever that may be. That would be a great waste of effort.[6]

The HGP would not alter its course.

Proposals to expand the EST project met with little or no support from funding institutions. The generation of fragments of random DNA sequence clashed with the dominant scientific ethic, that of dedication toward curing particular diseases or unraveling specific biological systems. It also did not match the profile of most pharmaceutical projects or biotechnology start-ups which typically had business plans structured around fulfilling particular unmet medical needs. Although Venter's initial studies were broadly directed towards neurobiology, they still resulted in a flood of As, Ts, Cs, and Gs, a distracting tangle of data that, at first glance, revealed little about neurological function or dysfunction. Venter was nonetheless driven to carry through on his gene discovery scheme, and soon he and several colleagues found support to pursue the project without the NIH.

Venter wanted to continue to act in the interest of science, to be part of the scientific community, sharing and cooperating in the pursuit of knowledge. He had no desire to run a business. So venture capitalist Wallace Steinberg, acting on behalf of the Health-Care Investment Corporation, set up a rather unusual enterprise designed to accommodate Venter's egalitarianism and at the same time provide investors with the rewards they sought. Venter and Steinberg had overlapping interests in wishing to speed up the discovery of human genes. They believed that doing so would both advance science and hasten the production of new drugs. Under the novel arrangement, Venter and colleagues would apply the EST approach to gene discovery at a nonprofit institute that would give preferential access to the results to a for-profit company. The Institute for Genome Research (TIGR) and Human Genome Sciences Inc. (HGS) were thus established in 1992, with Venter placed in charge of TIGR and cancer specialist William Haseltine appointed to run HGS. HGS promised to fund TIGR with $85 million over 10 years in exchange for marketing rights to the institute's discoveries.[7]

Far from the high-profile ruckus in Maryland (home of the NIH, HGS, and TIGR), a small company in California began pursuing a similar cloning and sequencing strategy. Whereas HGS and TIGR were born from a fresh mold and in the backyard of an indignant scientific community, Incyte Genomics (then called Incyte Pharmaceuticals) emerged quietly out of the biotechnology industry's school of hard knocks and into the crucible of a roaring computer industry. The company was incorporated in April of 1991 from the ruins of the biotech firm Invitron Inc. and its subsidiary Ideon Inc. Initially, Incyte's main mission was to take proteins with hints of desirable properties and make them into disease treatments, for example, bactericidal/permeability-increasing factor for sepsis, and protease nexin-1 for chronic inflammation.

It was an approach that multitudes of other biotech companies were trying—at an extremely low rate of success. For every erythropoeitin or insulin there were dozens of other proteins which, though they brought a wealth of biological knowledge, did not yield new drugs. Behind the handful of glamorous success stories was a wasteland of biotech shops angst-iously burning up their investors' cash like it was kindling under a pile of wet logs. Sepsis, in particular, has been like quicksand for biotech companies. The condition, a toxic response to bacterial infection common among hospital patients, defied numerous rational attempts to tame it. Besides taking the lives of its helpless victims, the disease left would-be rescuers flailing. Hundreds of jobs were lost and entire companies went under following unsuccessful pursuits for a cure for sepsis. Incyte would avoid such a fate.

Incyte's rise from the biotech muck was largely due to the actions of Randy Scott, an Incyte cofounder and its chief scientific officer. Scott had been trained as a protein chemist at the University of Kansas, where as a graduate student he was known for his grandiose schemes. His visionary ideas were restrained in two postgraduate research jobs, before finally taking wing at Incyte. There, Scott and colleagues sought to identify proteins made by a particular group of infection-fighting cells. Among the thousands of proteins produced by these cells a handful were anticipated to have properties that would render them to be suitable drug candidates. Scott figured that one could easily access and learn about these molecules through a crude purification procedure followed by partial sequencing of randomly selected proteins. When this was accomplished the resulting amino acid sequences indicated that, in addition to some well-known and expected proteins, there were quite a few novel proteins (sequences that had not been catalogued in GenBank or any other public databases). Some of these novel proteins appeared to have sequence similarity to known proteins, while others were entirely unrelated. Scott and colleagues wanted to identify all these proteins, measure their relative abundance, and understand their function in fighting infection. In 1991 biologists still knew less than 2 percent of the molecules produced in any tissue or cells being studied. Drug developers were therefore handicapped, as were physicians, who were just beginning to understand and rationally manage the molecular events underlying patient illnesses. Scott imagined that Incyte could provide the rudimentary information that would enable molecular medicine to flourish.

The Venter group's EST work made the front page of *The New York Times* in 1991, and Scott read it with great interest. He realized that at that time a cDNA-based approach held many advantages over protein-based approaches. Besides being less technically challenging than a similar protein-based operation, large-scale cDNA sequencing would produce cloned cDNAs, which are generally much more useful than crudely purified protein fragments. In addition, one could deduce protein sequence from cDNA, whereas redundancy in the genetic code made it impossible to ascertain DNA sequences from protein sequences.

Scott wanted to use large-scale cDNA sequencing to create a catalogue of human gene sequences. He shared the idea with Incyte's chief executive officer, Roy Whitfield, a former management consultant and an experienced biotech executive. The Scott-Whitfield partnership would show enduring success. They knew that Incyte would have to move quickly. Scott secured the support of Incyte's dozen or so employees, and a handful of them began preparing cDNA libraries and sequencing clones. Later, Scott and Whitfield won over Incyte's board of directors and helped attract additional financial support. The company held an initial public offering in November of 1993, raising about $17 million. By then, Scott, a son of a preacher, had become adept at articulating a vision of how ESTs could form the basis for a dramatically new approach to biomedical research. His vision would capture the imaginations of the scientific directors of the large international pharmaceutical companies that dominated the drug business.[8] Executives from Pfizer Inc., an American powerhouse among the mostly European drug conglomerates, arrived at Incyte's puny Palo Alto lab in a stretch limousine. They became as impressed with Incyte's sequencing plan as the Incyte group was with their drug business savoir faire. Incyte would soon be transformed from rags to riches, a not uncommon occurrence in Palo Alto, where Apple Computer and a host of other successful companies had their humble beginnings.

Whitfield and Scott showed their own remarkable business savoir faire when they presented a gutsy new business model to investors and prospective pharmaceutical customers/partners. Whereas virtually all other biotechnology companies made exclusive deals with the drugmakers, Incyte insisted on and ultimately received nonexclusivity. They would sell a common set of EST-based information over and over again. In addition, they would negotiate and receive reach-through rights, the promise of a small-percentage royalty on any drugs made from discoveries in Incyte databases. Incyte would be responsible for securing patent protection, first on the ESTs and then later on the full-length gene sequences.

Pharmaceutical companies normally would never consider giving away rights to future profits just for bits (and bytes) of information, particularly not for nonexclusive access to information, and most particularly not for nonexclusive access to information that is hundreds of millions of dollars (in preclinical and clinical studies) away from a drug product. What would compel them to act so differently with Incyte? First, the terms were reasonable. The royalty rate, for example, was a percentage in the low single digits. Second, no one knew for sure which genes would lead to new drugs and the winners in the race to both find human genes and secure patent rights would not be clear for years to come. Incyte's bid to rapidly sequence all human genes stood a good chance of succeeding, and a subscription to Incyte's sequence database could prevent a pharmaceutical company from being excluded from a significant set of intellectual property. Third, although the first sets of sequences were provided on a simple spreadsheet, Incyte was very rapidly becoming a leader in applying computer systems to capture, evaluate, and display huge amounts of biological data. A deal with Incyte would put a drug company at the cutting edge of this new trend. At least two drug companies made equity investments in Incyte in addition to buying subscriptions to its nonexclusive database.

Incyte positioned itself to be a dominant "tool" supplier for drugmakers, as Levi Strauss did for miners a century ago. When large quantities of gold were discovered in the foothills of California's Sierra Nevada Mountains in 1848, thousands of fortune-seekers flocked to the area. Only a tiny fraction of them struck it rich. However, companies such as Levi Strauss were able to establish profitable and enduring businesses by providing essential equipment (and blue jeans) to the miners. Incyte sought to establish itself as a supplier of gene information and gene analysis software to companies skilled in other aspects of drug development, manufacturing, and sales. If such externally controlled "tools" were to become critical components in the drug development process, then this would be a significant, if not revolutionary, point of departure for the pharmaceutical industry.

Drug companies, many of which were founded in the nineteenth century, were still vertically integrated. This meant that they usually had dominion over all aspects of the drug development process, from research in drug target discovery and medicinal chemistry to preclinical experiments, toxicological testing, and clinical studies (human drug trials). Computer companies had also been vertically integrated at one time (see Fig. 7.1). In the 1970s and 1980s Unisys, Amdahl, IBM, Digital, Commodore, and Apple practically built their computer systems from scratch (or they used exclusive suppliers), but by the 1990s that had all changed.

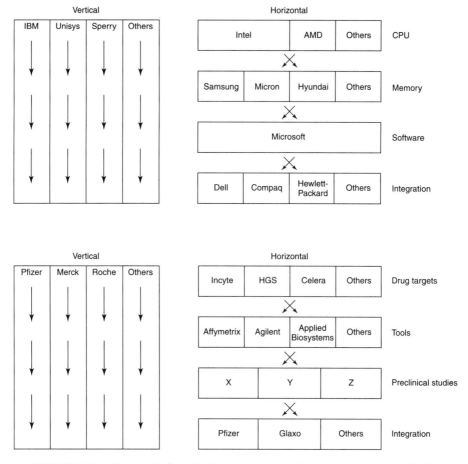

FIGURE 7.1 The work flow in the computer industry was once vertically integrated within single companies. Now it is horizontally integrated. The pharmaceutical industry may undergo a similar transformation.

Intel made most of the central processing units, Microsoft made most of the software, and a variety of different companies produced the disk drives, memory chips, monitors, and other components. Finally an end-manufacturer slapped them all together and marketed the computers as Dell, IBM, Compaq, Hewlett-Packard, Gateway, and so forth. This horizontal stratification probably helped the computer industry achieve greater efficiency, lower prices, and more rapid advances. Perhaps the pharmaceutical industry, mostly a collection of conservative old European firms, was finally ripe for such a makeover.

A large number of people, though certainly not all, seek better medicines to treat their ailments, particularly medicines that meet certain scientific standards of safety and effectiveness. Pharmaceutical companies seek to fulfill these needs through new and improved drugs. They also continually face revenue voids each time the 20-year patent life of one of their moneymaking drugs expires. Could such economic forces drive the development of new technologies? In 1993 Human Genome Sciences announced a $125 million deal with SmithKline Beecham Inc. (exclusive of others, at least initially). In 1994 Pfizer and then Upjohn Inc. bought nonexclusive subscriptions to Incyte's database in deals worth $25 million and $20 million, respectively. The EST scheme had been further validated. Later HGS would amend its agreement with SmithKline Beecham and then sign up two additional pharmaceutical partners, while Incyte would go on to sell database subscriptions to more than half of the 20 largest drug companies.

What about advancing the science of biology? Commercial EST shops appeared to be pimping random assortments of genes for cash. What could be more crass? The process was driven not primarily by the pursuit of knowledge but rather by business principles: higher throughput, lower costs, economies of scale, and nimbleness and flexibility in the marketplace. It seemed to fulfill at least some of the pharmaceutical industry's needs, but what did it all have to do with biology? What larger impact could the EST scheme have on the life sciences? Could economic forces, the subject of a "dismal" science, also propel a new approach to the study of life?

8

ESTs, the Gene Race, and Moore's Law

Incyte's and Human Genome Sciences' EST efforts raised considerable concern among academic researchers and those pharmaceutical companies that found themselves without access to a rapidly growing body of fundamental biological information. At the outset Venter and other TIGR officials had stated that TIGR sequences would be publicly available. Many scientists interpreted this to mean that like the prior NIH-supported data set the HGS-TIGR sequences would enter the public domain. Grant money was denied to new proposals for government-funded large-scale EST programs, reportedly because of the presumed redundancy.[1]

But by 1994 the naiveté of academic and government scientists finally met the harsh reality of the business world. Human Genome Sciences owned the rights to TIGR sequences. TIGR's sequence data *were* available to all academics, that is all academics who would sign away rights to future discoveries derived from the EST data (the dreaded reach-through rights). Academic and government research institutions were renowned for being fountainheads of innovation and discovery. They had recently become accustomed to out-licensing the discoveries of their scholars and scientists, but they were not accustomed to in-licensing the intellectual property of others, particularly not property that consists solely of information. They were not willing to grant reach-through rights.

Thus for a couple of years EST data remained inaccessible to most researchers. Finally, Merck & Co., then the largest American drug company, made a bold strategic move. In September of 1994 the company announced funding of an EST program to be carried out by a consortium centered at the Washington University in Saint Louis.[2] The cDNA libraries would be constructed in Professor Bento Soares' lab at Columbia University and arrayed for high-throughput processing by Greg Lennon, a scientist at the Lawrence Livermore National Laboratory. Over 4000 sequences a week would be dumped into GenBank with absolutely no restrictions on their use. Furthermore, unlike the TIGR sequences, there would be no privileged or advance access to the data. Alan Williamson, a vice president at Merck, proclaimed,

> Merck's approach is the most efficient way to encourage progress in genomics research and its commercial applications: by giving all research workers unrestricted access to the resources of the Merck Gene Index, the probability of discovery will increase. The basic knowledge thus gained will lead ultimately to new therapeutics for a wide range of disease—while providing opportunities—and preserving incentives—for investment in future gene-based product development.[3]

What might have motivated a for-profit company to spend its resources on a public database of information? Why would a lioness hunt down prey and then walk away, leaving her cubs to clamor over the carcass among all the other hungry beasts?

In one sense, by providing the Merck Gene Index, its term for the resulting data set, Merck was upholding its historical position as a respected leader in promoting health care and science. The company publishes *The Merck Index*, a one-volume encyclopedia of chemicals, drugs, and biologicals that appears on almost every chemist's bookshelf. It also produces *The Merck Manual*, which is said to be the most widely used medical text in the world, providing useful clinical information to practicing physicians, medical students, interns, residents, and other health care professionals since 1899.[4]

Merck's EST initiative was also a shrewd patent-neutralizing strategy, for it could prevent competitors from obtaining rights to genes that Merck or anyone else wished to develop into new drugs or diagnostics. Once an invention has been disclosed, then any new patent applications on it must be made within a year or the patent office will not even consider them. Thus would-be gene patentees must find patentworthy sequences prior to others *and* file patent applications on them before they have been in the public domain for a year. It would be tougher to fulfill these requirements with Merck/Washington University sequences pouring into GenBank. If it turned out that partial gene sequences were to be the basis for gene patents, then genes that were first represented by Merck/Washington University ESTs or had been sitting in GenBank for a year could not be patented. Or, if the full-length gene held its ground as the smallest patentworthy unit, then Merck/Washington University ESTs could provide an army of academic researchers, as well as others, with a greater opportunity to beat commercial EST enterprises at deriving full-length genes and obtaining gene patents.

The first release of the Merck-funded project was made in February of 1995. The Merck/Washington University EST sequences soon accounted for more than half the sequence records in GenBank and became the basis for a subsection of the database known as dbEST (database of ESTs). The data was made queriable, and anonymous downloads via the World Wide Web were enabled.[5] Everyone with an Internet connection had an equal chance to begin to find interesting ESTs, develop insights into gene functions, and discover new drug and diagnostic leads (www.ncbi.nlm.nih.gov/). Even the physical clones were available, at minimal cost and with few restrictions. By 1998 sequences were being added at a rate of over 1500 per day.[6]

Gene-focused biologists were already overjoyed with GenBank and with their own burgeoning sequencing capabilities. Access to an EST database was icing on the cake. Eager researchers searched the EST databases often, and, as anticipated by early EST proponents, they quickly found interesting ESTs and developed them into noteworthy discoveries. ESTs hastened the discovery of genes associated with colon cancer, prostate cancer, Alzheimer's disease, and numerous other diseases and traits.[7] Often the short sequences contained enough information to suggest that they encoded a portion of a particularly interesting protein. An EST containing a characteristic "death domain" sequence came from a novel gene which, after several months of laboratory work, was found to have tumor-cell-killing activity, as was suggested by the "death domain."[8] Hundreds or perhaps thousands of such findings were made in the 1990s, although because the functional analysis of these genes is still quite slow the full impact of these EST discoveries won't be known for years to come.

The full impact of gene patents also remains to be seen. By 1999 dozens of full-length genes were receiving patent protection each month, and the total number of gene patents pending was well over a thousand (as revealed through international filing disclosures). Since these patents and patent applications are from numerous institutions, both nonprofit and for-profit, it is improbable that any one group will own 20-year rights to all human genes or even to a majority of all genes. The Merck/Washington University initiative helped to level the playing field. ESTs seem to be providing gene bounties for all that are interested.

The resulting scientific achievements would stimulate financial support for further EST production, which in turn would lead to more scientific achievements. Business would affirm the science and science would affirm the business in an upwardly spiraling exchange. Furnaces were stoked and the sequencing factories blazed ahead, some operating 24 hours a day. "Faster, cheaper, and better"—a mantra that captures the fervor of change in the computer industry—was chanted by molecular biologists. The mantra has even been enacted into a "law," Moore's Law, which states that every 18 months the processing power of the most widely used computer chips will double, with a concomitant decrease in cost. Intel Inc. cofounder Gordon Moore made this observation in 1975, and to the astonishment of many it has held true through several decades. Failure to adhere to the law is punishable by obsolescence.

Moore's Law has been alluded to so often and for so many aspects of the computer industry that it has become a droll cliché. Nonetheless, in the mid-1990s the concept was still quite new to molecular biology, and was embraced perhaps most ardently by the management of Incyte, who happened to work in Silicon Valley. Incyte made sequence output a high priority, ensuring that their sequencing efforts obeyed Moore's Law. Technology advances drove this self-fulfilling prophecy, with automation introduced in almost all aspects of cDNA preparation and sequencing. Robots picked the bacterial colonies and machines extracted DNA from them. Additional innovations squeezed more sequence from fewer materials, and the enormous scale of Incyte's operation reduced their per base pair costs even further.

The processing, tracking, storing, and displaying of information are the quintessential tasks of the computer sciences. The industrial-scale sequencing of the entire set of human genes would summon a wide range of new technologies and computational tools that in turn would propel a revolutionary new approach to the understanding of life. The drive to make things "faster, cheaper, and better" is really a mantra of the entire Industrial Age. Why shouldn't it be applied to the medical sciences and to the information necessary to understand the molecular basis of life?

The ultimate scientific goal for each of the large-scale sequencing projects is knowledge of all human genes, or *closure*, as it came to be known. Closure can take many forms, including: ESTs representing all expressed genes, the full protein-coding sequence of all genes, the genomic sequence of all genes (which would include introns and regulatory sequences), physical clones for all expressed genes, or knowledge of the function and regulation of all human genes and their variants (polymorphisms).

In 1998, with about 5 million ESTs in the largest database and about 2 million ESTs in the public domain, one might imagine that the entire set of expressed human genes would be represented, whether they numbered 40,000 or 400,000. However, it could be shown that the EST databases were not complete, for some genes that had been discovered through traditional directed cloning techniques were not represented in the EST databases. One study showed that about 6 percent (6 of 94) of the genes that either suppress or stimulate cancer growth[9] and 9 percent (8 of 91) of positionally cloned genes were not represented in dbEST.[10] Partial sequencing operations continued at Washington University, Incyte, and a host of newer outfits, including GENSET in France,

Millennium Pharmaceuticals, Hyseq Inc., and AXYS Pharmaceuticals Inc. There were several reasons to continue sequencing, chief among them being that new genes were still being discovered and additional information was being obtained on previously tagged genes through additional partial gene sequences.

With the help of ESTs the race to identify and sequence all human genes seemed to be near completion, but no one could tell how near. How many human genes had been identified, and how many total human genes are there? The EST data sets provide an important resource for tackling the total gene number question, but, for reasons that will soon be made clear, they are not likely to provide a definitive answer. One should, however, be able to estimate the number of genes represented by the set of known ESTs. A group at the NCBI collapsed dbEST into a nonredundant set, which they call the *UniGene* set. In 1998 Uni-Gene included 43,000 sets of overlapping gene sequence fragments, termed *clusters*. If each cluster of ESTs represented a unique gene then there were at least 43,000 genes represented in dbEST. The TIGR EST database was determined to contain approximately 73,000 genes.[11] However, the same set of ESTs can be collapsed or assembled in different ways and thus may provide different solutions to the question. Even after significantly more sequence information became available gene number estimates varied widely, ranging from 35,000 to about 150,000.[12] Obtaining definitive total gene counts from EST databases has proved to be rather frustrating, although the process does illustrate some important computational issues.

Messenger RNA sequences conclude with a string of "A" nucleotides and are captured using a string of "T" nucleotides, a kind of molecular Velcro. Sequencing of the subsequent cDNA may be done from either the downstream end where the As are (known as the 3-prime or 3′ end) or from the upstream end (known as the 5-prime or 5′ end). Since mRNA tends to degrade rapidly, particularly from the 5′ end, many cDNAs are incomplete, missing varying degrees of sequence from the 5′ end. Therefore, 5′ ESTs from the same gene may cover different portions of the gene. It is still possible to re-create the entire gene sequence from these fragments by assembling together overlapping sequences and generating a consensus. The process is analogous to combining the following group of sentence fragments (1) ["be assembled from a collection of ESTs."] (2) ["In this way, a large portion of a gene, in fact often"] (3) ["gene, in fact often the entire protein coding sequence, can be assembled"] thusly: "In this way, a large portion of a gene, in fact often the entire protein coding sequence, can be assembled from a collection of ESTs."

Such EST assembly is one of the earliest applications of *in silico* biology. For although each EST was from a cDNA that was sequenced in a plastic or glass (*in vitro*) tube, the entire coding sequence of the gene was found electronically (*in silico*) from an unanticipated set of ESTs that may well have been derived from distinct tissues and from distinct individuals. Now, consider what happens when more and more ESTs, derived from various types of cells, are entered into the database. Initially, the number of clusters overlapping sets of genes should increase as more genes become represented. Eventually, the number of clusters will likely exceed the total number of genes, for there will be more than one cluster for many genes (see Fig. 8.1). Then, as clusters are joined by overlapping ESTs, the total number of clusters should diminish and stabilize at the total number of expressed genes. The number of *singletons*, ESTs that fail to cluster, should rise and then fall to zero. However, in practice the number of singletons never stopped rising! New genes were indeed being discovered, but there could not be an infinite number of genes as the near-constant slope indicated. Something was amiss.

It turns out that a host of biological and technological factors complicate cluster analysis of EST databases. On the technology side, the cDNA libraries used in sequencing had contaminating sequences from nonhuman organisms, genomic DNA, and from immature RNAs that had yet to have all their exons spliced out. Known contaminant sequences, such as those of *E. coli*, can be simply deleted from the database. (*E. coli* bacteria are used in production of the cDNA libraries. Fortunately, their entire genome is known, so it can be electronically deleted.) The presence of many known contaminant sequences suggests that unknown ones are

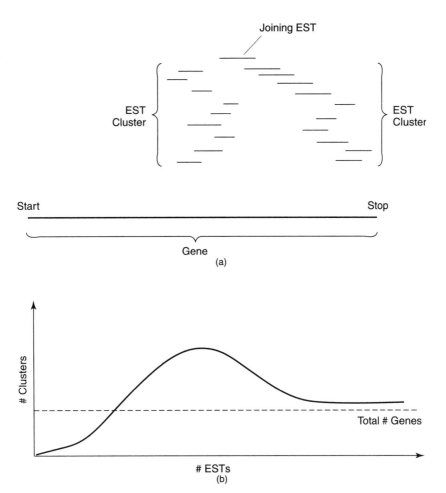

FIGURE 8.1 (a) An unknown gene sequence is depicted below a series of partial sequences (ESTs). Groups of two or more overlapping ESTs form clusters. A joining EST unites two previously independent clusters. Prior to their joining, one might be led to believe that the clusters come from two distinct genes. (b) The number of clusters rises as the number of ESTs rise, until the occurrence of joining ESTs exceeds the number of new clusters formed. Then the clusters "collapse," and the number of clusters approaches the total number of genes.

likely to lurk within the EST databases. Unknown sequences are thus not just unknown, they are also of uncertain origin.

In addition, sequencing machines, particularly ones optimized for high-throughput, typically churn out sequence data in which 1 to 3 percent of the nucleotides are incorrect. They also may inadvertently join two unrelated sequences. Furthermore, clustering algorithms can be tripped up by sequences that are nonidentical but highly similar and by naturally occurring splice variants, mRNAs that are the result of alternative cutting and pasting of RNA (in a literal sense). Last, although the enzyme complexes that replicate DNA and make RNA from DNA are remarkable in their ability to start and stop at specific locations, they are not 100 percent precise. In sampling millions of RNAs, even small transgressions could enter the databases.

Thus it is no small challenge to build an algorithm that can distinguish between sequence variations caused by technological glitches and those due to glitches (or features) inherent to the molecular design of life. Fortunately, the data set is big enough and "clean" enough for some new and sophisticated algorithms to create useful clusterings of ESTs. Besides enabling the assembly of full sequences from ESTs, clustering algorithms identify a set of nonredundant (or nearly nonredundant) clones that can be used for further experimentation, as will be discussed in later chapters.

Sequencing randomly selected clones proved to be a great way to gain access to the most commonly expressed genes. But the gene discovery rewards for such EST sequencing diminishes in time, and at some point the payoffs may no longer justify the effort. To illustrate the problems with random sampling, consider the following: Suppose that you are at a carnival and there is a large barrel filled with 1000 individually numbered balls. One of the balls leads to a prize. You are allowed to pick five balls at a time. However, after examining the numbers you must put the balls back into the barrel. After examining 200 handfuls (1000 balls) you may be no closer to finding the prize ball then when you started. It is like finding a needle in a haystack, only worse. For only a small portion of the haystack is examined at one time, and after each look the entire haystack is re-created. Such is the nature of random sampling.

Numerous tricks have been applied to overcome the limitations inherent to random sampling strategies. Microdissection and other sampling techniques have allowed greater representation of genes whose expression is restricted to a small group of cells or to a narrow portion of the human lifespan. And to increase the representation of rarely expressed genes in a cDNA library, the cDNAs may be denatured and allowed to reanneal. Those DNA strands that find partners readily can then be eliminated. The relative abundance of rarely expressed genes is thereby increased in the resulting, reconfigured cDNA library.[13] Just as advances in technology have allowed more gold to be mined from old claims and more oil to be extracted from once-abandoned wells, innovative laboratory procedures have enabled the discovery of rare or otherwise elusive genes through EST-based approaches.

Despite the success of these gene mining techniques, EST sequencing efforts eventually ran up against the issue of diminishing returns, and efficiency became of increasing concern for the operators of the EST sequencing factories. Moreover, EST sequencing could not access the nonexpressed sequences in the genome, and so gene regulatory regions in particular remained untapped. So ultimately, in the end game of the contest to find and characterize all human genes, the tables turned and random EST sequencing proponents, first Craig Venter and later Randy Scott and others, considered genomic DNA sequencing.

9

The End Game

The year 1998 marked the beginning of the end of the DNA sequencing race. In that year two new companies set themselves up to sequence the entire human genome; the first animal genome was fully sequenced; and Human Genome Project participants greatly increased their sequencing output and advanced their timeline for completion of the human genome. In addition, it was in 1998 that the first product from an EST-derived gene made its way into clinical trials.

As the race to access the information content of the human genome entered its final stage, the various contestants for the prize differentiated themselves by their end-game strategies. As early as 1993 Human Genome Sciences Inc. (HGS) had claimed to have ESTs representing 70 percent of all human genes, and a few years later they had allegedly reached the 95 percent mark. Apparently, this was enough, for the company then turned its attention away from human EST sequencing and toward drug development.

HGS's strategy was to select some of their ripest fruit, the gene sequences that were most likely to encode proteins that are active in the bloodstream, and use them as a springboard for becoming a full-fledged pharmaceutical company.[1] At that time cell-secreted proteins (those that are secreted from the cells in which they are made) had been the bedrock of almost every profitable biotech company. Amgen's Neupogen® (for cancer treatment), Genentech's Activase® (for strokes), and Biogen's Avonex® (for multiple sclerosis) were formulated from the secreted proteins granulocyte colony-stimulating factor, tissue plasminogen activator, and interferon beta-1a, respectively. Proteins may be secreted out of cells by virtue of a particular pattern of amino acids that appears in the first 20 positions of their amino acid sequence. These amino acids route the protein to the cell's outer membrane, in a manner that is akin to a zip code routing letters to a particular post office. About 15 percent of all genes encode cell-secreted proteins, and these genes can be identified by a characteristic sequence of 60 base pairs. If a newly identified sequence encodes the secretion zip code, then it is almost certain to be of a cell-secreted protein gene. Genes that do not encode cell-secreted proteins (proteins that remain within cells) may be good targets for medicinal chemicals, but they require more resources to develop, and thus HGS decided to leave them for their large pharmaceutical company partners.[2]

Preclinical (animal) studies had shown that HGS's Myeloid Progenitor Inhibitory Factor-1 (MPIF-1; a secreted protein) was able to protect bone marrow cells from the harmful effects of chemotherapy. Then in December of 1997, HGS received clearance from the FDA to initiate clinical trials with MPIF-1—making it the first product from an EST-derived gene to reach this milestone. HGS was eager to focus on MPIF-1 and take it all the way, since its appetite for finding new human sequences had already diminished considerably by this time.

In June of 1997 TIGR and HGS had broken their contractual agreement prematurely. The five-year, highly successful relationship between the nonprofit institute and the for-profit company was terminated. No longer would HGS underwrite TIGR's sequencing operation in exchange for the exclusive rights to ESTs. After the breakup, HGS was left in possession of more than 900,000 ESTs, but TIGR would no longer be funneling them new sequences.

TIGR was a nonprofit institute, and its mission had always been to provide sequence information to all researchers. The divorce from HGS helped free the release of 110,000 human ESTs. This set joined all the liberated sequences in GenBank, where several other sets of TIGR sequences also resided, having been funded by less restrictive sources. Sequencing projects continued at TIGR without HGS; there was even one to sequence human genomic DNA for the Human Genome Project.

TIGR, with Craig Venter at the helm, had succeeded in being the first group to ever sequence the entire genome of a free-living (nonviral) organism. In 1995 a team of about 30 TIGR researchers in collaboration with scientists at Johns Hopkins University, the State University of New York at Buffalo, and the National Institute for Standards and Technology, published the sequence of all 1.8 million base-pairs of a harmless strain of *Haemophilus influenzae*, a bacteria that can cause respiratory infections and meningitis.[3] To delineate the bug's DNA sequence, Venter and crew used a "shotgun" sequencing approach, in which the *Haemophilus* genome was randomly sheared into millions of different fragments, the fragments were cloned, randomly selected clones were sequenced, and a computer program then assembled overlapping sequence fragments into one enormous contiguous sequence. The technical similarities with the EST project should be apparent. Also, like the EST plan, the idea had earlier roots. Fred Blattner of the University of Wisconsin at Madison, for example, had suggested back in the late 1980s that shotgun sequencing methods could be applied to both the entire human genome and genome of the *E. coli* bacterium, the focus of his work. However, at that time it was impossible to assemble sufficient resources for such a project.

A shotgun approach had never been used on a piece of DNA as large as the *H. influenzae* genome. Lengthy pieces of DNA had been sequenced, including entire chromosomes from yeast, but only by way of tiled sets of fragments, overlapping and preordered pieces of DNA. Venter and all had calculated that the sequencing of 1.8 million base pairs of randomly selected DNA fragments would leave 37 percent of the *H. influenzae* genome unsequenced (from Poisson's equation $P_O = e^{-m}$, where P_O is the probability that a base pair will not be sequenced and m is the fold coverage of the full sequence). With 11 million base pairs sequenced, enough to sequence any particular base pair an average of six times over (sixfold coverage), the probability of not sequencing a particular base pair would be reduced to 0.25 percent, still enough to ruin the possibility of electronically reconstructing the available sequences in their proper order. Nonetheless, the team found that by shearing the bacterial genome into two groups of fragments, one of about 2000 base pairs and the other of from 15,000 to 20,000 base pairs, and by partially sequencing 500 base pairs from both ends of these cloned fragments, gaps in the final sequence would be minimized and could be readily identified and sequenced. Thus, in this case there appeared to be a solution to the random-picking problem.

Although the whole-genome shotgun approach required more raw sequencing than the tiling approach, it had one great advantage. It eliminated the need for the comparatively slow and costly process of mapping and ordering gene fragments prior to sequencing. The successful release of the *Haemophilus* data convinced everyone of the utility of shotgun sequencing on small genomes. It also generated significant scientific and medical interest, in part because many strains of bacteria were resistant to known antibiotics and posed a significant health problem that could be addressed in new ways if bacterial sequences were available. In fact, the *Haemophilus* achievement proved to be a great inspiration. Incyte, Genome Therapeutics Inc., the Sanger Centre in England, several groups of NIH-supported researchers, and other teams thereafter utilized shotgun sequencing strategies to fully sequence the genomes of a wide variety of bacteria, while TIGR itself managed to complete eight more genomes by 1998.

Remarkably, Venter had applied for NIH funding for the *Haemophilus* project, but had been turned down, reportedly due to doubts about the shotgun sequencing approach.[4] Instead, grants from HGS and the American Cancer Society helped support the successful effort. Upon release of the *Haemophilus* sequence, Venter again caused a ruckus in the human genome contest. He was quoted by *Science* magazine as saying that the *Haemophilus* work had "raised the ante worldwide for sequencing the human genome."[5] Venter asserted that the random shotgun approach could be used on the entire genome of *Homo sapiens*, a collection of DNAs that is about 1600 times the size of the *Haemophilus influenza* genome. Unmoved by Venter's proposal, the administrators of the Human Genome Project held fast to a course that increasingly diverged from that of TIGR and its controversial CEO. The HGP marched on with a budget in the United States of $243 million, $267 million, and $303 million in 1996, 1997, and 1998, respectively.[6] It funded seven major sequencing centers in the United States. In addition, the Wellcome Trust foundation helped to establish the Sanger Centre in England as a major contributor to the Human Genome Project, and programs in several other countries were also contributing.

In 1997, near the halfway point in the 15-year project, numerous accomplishments could already be seen, particularly in generating a map of the human genome. About 8000 markers, sequences that often show variation from one individual to the next, had been located throughout the genome.[7] ESTs from about 15,000 different genes were also mapped.[8] These markers and mapped genes aided in the discovery of numerous disease-associated genes. They also served as guideposts in creating and tracking a growing set of overlapping DNA fragments of 40,000 and 400,000 base pairs in length. The fragments were maintained and manipulated in a form known as the *bacterial artificial chromosome*, or BAC. Eventually, 30,000 BACs would be sequenced in their entirety.

Ordering and aligning hundreds of thousands of these fragments was a tremendous challenge. Fragments were identified by the markers they possess and by the other fragments with which they overlap. Sets of cloned sequences that cover the greatest region of the chromosome but which have minimal overlap (the minimal tile set) were selected for sequencing (see Fig. 9.1). Sequencing itself was intended to lag behind mapping and technology development efforts, and by the end of 1997 barely 60 million base pairs had been completed, approximately 2 percent of the human genome.[9] At that time there was some concern about meeting the year 2005 deadline. Scientists at one sequencing center wrote that: "To complete the genome by 2005, starting in 1998, seven large-scale sequencing centers, for example, would each have to complete on the order of 75 Mb/year. Sequencing centers now have a throughput of 2 to 30 Mb/year. If the genome is to be sequenced on time and within budget, sequencing must become significantly faster and cheaper."[10] However, despite the formidable task that lay ahead, at the halfway point HGP participants were mostly optimistic that the entire genome would be sequenced by the year 2005. The entire genome of a species of yeast was completely sequenced in 1996 (all 12,067,280 base pairs), and sequencing of the first animal genome, that of the tiny worm *Caenorhabditis elegans,* was well on its way. (Nearly all 97 million or so base pairs would be completely revealed by the end of 1998.)

The HGP, its administrators, and its army of grant reviewers were open to new ideas; they funded research in many new and uncertain technologies. However, they were less open to alternative visions of the final masterpiece, the three-billion or so base pairs

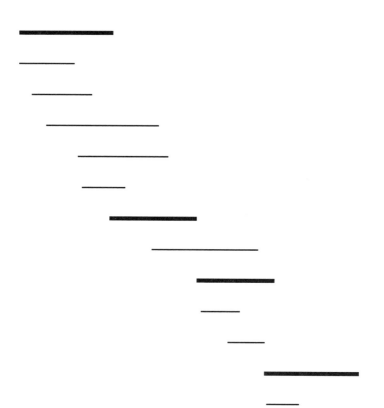

FIGURE 9.1   A collection of random large genomic fragments (BACs) is aligned. The optimal set of fragments for sequencing efficiency (the minimal tile set, indicated here by the bold fragments) consists of those that fully cover a region of the genome and have the least amount of overlap.

that would signify the accomplishment. HGP was committed to leaving almost no gaps and having a sequencing error rate of 0.01 percent or less. Wavering from this goal, it was thought, would put them on a slippery slope that would allow the final product to have patches of error-filled or, worse yet, incorrectly positioned sequences. This could greatly compromise the use of the final product, and it would desecrate the human genome. There was no assurance that shotgun sequencing the entire human genome would meet the requirements, and the idea had already been debated. In the journal *Genome Research*, James Weber of the Marshfield Medical Research Foundation in Wisconsin and Eugene Myers of the University of Arizona put forth a proposal that called for 10-fold coverage.[11] In response Philip Green, a HGP-supported researcher at the University of Washington, asserted that:

> We simply do not know enough about the structure of the [human] genome at the sequence level, or of the biases inherent in clone libraries, to simulate the sequencing process adequately. . . . Nor does the success of a whole-genome shotgun approach with bacterial genomes provide any confidence whatsoever that the same approach would work with the human genome. The fundamental difficulty in assembly is dealing with repeats, and bacterial genomes have few of these.[12]

Repeated sequences, such as the Alu sequences, could trip up the computer programs that are designed to assemble random sequences into contiguous ones. Green also noted how difficult it would be to realign dozens of coordinated tile-based operations into a cohesive shotgun-based operation. Clearly, HGP participants had a great deal invested in the clone-by-clone, tiled approach, emotionally, intellectually, and perhaps financially as well. Thus, despite the fact that human genome sequencing had barely begun, no one was very surprised that the latest, *Haemophilus*-inspired whole-genome shotgun sequencing initiative failed to take root within the HGP. Would it surprise anyone if Venter took matters into his own hands? Considering the past events it shouldn't have. Nevertheless it did.

Craig Venter and the Applied Biosystems Division of Perkin-Elmer, Inc. (the company itself is now named Applied Biosystems) dropped the bombshell in May of 1998. They declared that they would sequence the entire human genome in three years for a total cost of between $200 million and $250 million, roughly one-tenth the projected cost of the HGP. This was exactly the kind of bold initiative that was typical of Venter. But who would back such a scheme, given that (1) for years Incyte, HGS, their pharmaceutical company partners, and a host of others had been cherry-picking the most commercially valuable genes and (2) the most direct competitor was a multigovernment-backed organization that had an established infrastructure, deep financial backing, and was in the process of ramping up its own assault on the human genome, an entity that many wished to claim for all humankind and deny to any profiteers?

Venter and associates would carry out this project at a newly formed company 80 percent owned by Applied Biosystems and 20 percent owned by private investors. Applied Biosystems is a leading manufacturer of sequencing machines. Their ABI Prism™ 377 Sequencer, which was introduced in 1994, had become a staple of sequencing operations everywhere. The human genome sequencing announcement was coupled to the unveiling of Applied Biosystems' new sequencing machine, which was said to be able to process about 1000 samples in 24 hours, while requiring only about 15 minutes of hands-on labor. Michael Hunkapillar, a key inventor of automated DNA sequencing technology and the president of Applied Biosystems' sequencing machine division, would join the new company, which was named Celera (from a root word that means swiftness) Genomics Corporation.

The founders of the new company outlined their plan, including their scientific, patent, and business strategies, in public talks and in an article published in *Science* magazine. Though it was certainly grand in scale, the scientific plan was otherwise unremarkable. It called for whole-genome shotgun sequencing with 10-fold coverage (an outstanding 70 million sequencing reactions, performed on 35 million cloned DNA fragments). A set of large fragments that had already been mapped would be used to help order thousands of partially assembled sequences. The founding scientists acknowledged that repetitive sequences could pose a problem for computational assembly, but suggested no new means for dealing with this issue.

Polymorphisms (also known as *gene variants*) are an inevitable by-product of the sequence assembly process. When overlapping sequences from different individuals are aligned, instances of a particular type of polymorphism, *single nucleotide variations* (SNPs), are readily observed. SNPs mark the places in the genome where individuals differ. Venter and crew intended to capture SNP data. Some of these SNPs, either separately or in combination, would likely be associated with traits and diseases. (A detailed description of human genetic variation and SNPs follows in chapters 11 and 16.) At the outset it was not at all clear which would be more valuable to Celera, the gene sequence data or the SNP data; however, by shotgun sequencing several human genomes the company could pursue both simultaneously.

To generate revenues Celera would sell products or services associated with "high-throughput contract sequencing, gene discovery, database services, and high-throughput polymorphism screening."[13] To protect its intellectual property, Celera would selectively apply for patents. "Once we have fully characterized important structures (including, for example, defining biological function), we expect to seek patent protection as appropriate . . . we would expect to focus our own biological research efforts on 100 to 300 novel gene systems from among the thousands of potential targets." Celera also planned to obtain patent protection on diagnostic tests based on the association of particular polymorphisms with important traits and diseases.

Venter had earned the respect of many for making ESTs into a successful vehicle for rapid gene discovery; for making the shotgun sequencing approach into a successful means of sequencing entire genomes; and for shepherding numerous bacterial genomes and other sequences through the TIGR sequence factory and into the public domain. On the other hand, Venter had outraged many folks by seeking patents on ESTs and by offering HGS exclusive rights to so many ESTs. Like a politician in office, Venter knew his constituencies' hot buttons. Patenting human DNA sequence and accessing human genomic sequence DNA were highly sensitive issues not only for scientists, but for real politicians and for the general public, as well. So, in presenting the Celera plan, Venter and partners knew that they must walk across a tightrope. Bravado and showmanship were carefully coupled with some look-me-in-the-eye American humble pie. "We are trying to do this not with an in-your-face kind of attitude," Venter said regarding the HGP. Homage was paid to the colossal project. The HGP was credited with "laying the groundwork for a revolution in medicine and biology." Furthermore, the Celera scientific team declared:

> It is our hope that this program is complementary to the broader scientific efforts to define and understand the information contained in our genome. It owes much to the efforts of the pioneers both in academia and government who conceived and initiated the HGP with the goal of providing this information as rapidly as possible to the international scientific community. The knowledge gained will be key to deciphering the genetic contribution to important human conditions and justifies expanded government investment in further understanding of the genome. We look forward to a mutually rewarding partnership between public and private institutions, which each have an important role in using the marvels of molecular biology for the benefit of all.[14]

Having declared that "it would be morally wrong to keep the data secret," Celera officials offered to publicly release portions of the sequencing data every three months and the complete genome at the end of the project.[15] To allay data quality concerns, the team set a goal of 99.9 percent sequence accuracy and offered to make the electropherograms [the rawest form of sequence data] available. The tightwire walkers also promised not to apply for patents on the vast majority of the sequences. They would seek to control only the most commercially promising novel gene sequences.

Naturally, the Celera announcements sent shock waves through the HGP group. It also evoked déjà vu, as it had all the elements of the EST sequencing controversy that had erupted seven years earlier, namely the patent and data accessibility concerns, the possible overlap with the public effort, and the differences in scientific strategies. The Wellcome Trust responded just three days after the Celera's initial announcement. The foundation declared that it would double its commitment to the HGP by supplying the Sanger Centre with an additional $180 million. Furthermore, Wellcome stated that in light of the recent developments, "The Trust is conducting an urgent review of the credibility and scope of patents based solely on DNA sequence. It is prepared to challenge such patents."[16] Celera's friendly overtures were apparently not enough to mollify this foundation.

In the United States the response was perhaps a bit more tactful, but no less intense. First of all, critics denounced the whole-genome shotgun sequencing scheme. They predicted that missing sequences and repetitive sequences would cause "catastrophic problems" in assembling sequence fragments.[17] They contended that the level of sequence accuracy would be insufficient. Others, however, were of the opinion that with Craig Venter leading the charge there was a good chance that the scheme would be made to work, and pointed out that the variations due to sequencing errors would be less significant than the variation between individuals.

People responsible for funding the HGP were compelled to re-examine the possibility that the $3 billion public effort might end up being redundant with a privately funded effort. Many questioned where all the money was going to and whether the project was worth the expense. William Haseltine of HGS went so far as to liken HGP to a "gravy train" for recipients of government funds. The U.S. House of Representatives subcommittee on energy and the environment held hearings on the matter. The current and future benefits of the HGP were highlighted, uncertainty over the success of whole-genome shotgun sequencing technique was noted, concerns over the quality of the method's sequence assemblies were discussed, and it was agreed that the HGP was on the right course. Ultimately, Francis Collins, the director of the HGP, declared that "the private and public genome sequencing efforts should not be seen as engaged in a race."[18] New funding was approved for the HGP. Thus, with Applied Biosystem's investment and the new HGP allocations, a new price of admission was paid and the human genome sequencing contest began in earnest.

Paradoxically, an intensified government-backed sequencing effort could help Celera's primary backer. Recall that Celera's initial announcement was coupled with the introduction of a new sequencing machine. Increased government funding could help Applied Biosystems sell more of its new $300,000 machines. Competing with a customer is generally not considered a good business practice, but clearly the HGP was a rather special customer, and clearly some extraordinary business forces were at work here.

Despite its careful posturing, the Celera group's May announcement was an "in-your-face" affront to Incyte. Incyte's stock dropped about 20 percent in the week after the announcement, and more than one investment analyst speculated that the heightened level of competition bode poorly for the Silicon Valley company. Then, in August of 1998 Incyte made a strategic move of its own, announcing its purchase of Hexagen, a SNP-discovery company based in Cambridge, England, which would become part of a separately financed business unit called Incyte Genetics. The new company's objective was to map the human genome within one year, sequence particular "gene-rich" regions of the human genomic DNA, and look for SNPs in every human gene. Incyte Genetics would seek $200 million over the first two years, which would come from its parent company as well as from strategic partners and additional outside investors. Thus, Incyte also anted up and was prepared to participate in the final leg of the race to gain knowledge of all human genes.

Back at HGP headquarters, just outside of Washington, D.C., team members were finalizing new and intensified plans. In October of 1998 Collins and colleagues presented "New Goals for the U.S. Human Genome Project: 1998–2003" in *Science* magazine. The report called for the complete human genome sequence by the year 2003, two years earlier than previously planned, and a lower-quality "draft" of the entire genome by 2001, the year that the Celera team was aiming for. They also announced several new goals for the HGP, including the identification of SNPs and the development of techniques that would allow gene functions to be determined at a rapid rate, thousands of genes at a time. The impetus for the new goals and the accelerated timetable was not explicitly given, though the authors offered the following thesis:

The HGP was initiated because its proponents believed the human sequence is such a precious scientific resource that it must be made totally and publicly available to all who want to use it. Only the wide availability of this unique resource will maximally stimulate the research that will eventually improve human health. Public funding of the HGP is predicated on the belief that public availability of the human sequence at the earliest possible time will lead to the greatest public good.[19]

It is easy to imagine where this might be headed. The next installment of this plan might call for further extension and expansion of the project. The drugs and treatments derived from the sequences might also be considered precious scientific resources . . . and so on. Could all of the downstream products of the HGP be brought into the public domain?

The journey towards knowledge of all our genes, from the early planning of the Human Genome Project in the 1980s to the development of EST factories in the early 1990s and new commitments toward achieving sequence closure in 1998, has been guided by no preestablished authority. In fact, the society that allowed this massive undertaking to progress shows all the hallmarks of freedom and anarchy.

Anarchy? Don't the participants live in democratic societies where people vote and elected officials are empowered to set policies? And don't they live under a capitalist system where the marketplace guides the pursuit of useful products? And aren't these lawful nations, where an enduring set of written codes establishes boundaries and rules?

Sure, but the hierarchically structured political systems and the hierarchically structured research organizations that they fund and oversee have not been the leaders in the quest for the human genome. They have been able to assemble an international team of highly ranked scientists and science administrators and secure hundreds of millions of fresh taxpayer dollars, but privately-funded ventures have beat them to the majority of human genes.

Nonetheless, the race to discover genes and gene functions can hardly be described as a free competition, as the consortium of governments and private foundations has been able to compete vigorously with private entities, yet is bound by an entirely different set of rules. It is not as if one group has encroached on the other's domain, such as a private business taking on the postal system, or Uncle Sam entering the auto business. Instead, in creating the science of genomics both the public and private efforts have feasted on ideas, information, and materials generated from each other.

For the newly hatched genomics industry, the laws of the land have offered little guidance. The journey has been more like a trip in a dark wilderness than a stroll along a well-marked pathway towards order. Proprietary rights to genes, gene fragments, and gene variants remain unclear. There have been a number of important legal rulings, but they have inevitably revolved around the state of science and technology that existed at the time of the invention in question (i.e., in the fast-receding past). With each new advance the applicability of these rulings comes into question, and thus the legal system remains a laggard.

Is anarchy an apt description of the social system in which genome discoveries have flourished? Not in the maligned meaning of the word—not disorder, chaos, or violence—but rather the original, placid, and simple meaning of anarchy, namely the absence of an authoritative structure. Anarchism was a political philosophy that vied with (and ultimately lost to) the ideas of Karl Marx at a time when people were ready to take on the gross inequities and injustices of early industrial societies. Much anarchist rhetoric, then and now, focuses on toppling existing power structures; hence the reputation for hostility and violence. However, quite a few thinkers, such as Pierre-Joseph Proudhon, Mikhail Bakunin, and Peter Kropotkin, have also contemplated peaceful societies without established power structures, and scores of anarchist communities have been created.

Under anarchy the power to make decisions remains firmly with individuals. Authority does not come from the will of a ruler, the majority, particular ancestors, or other historic figures. It need not be passed on to a representative once every few years, and it doesn't need to be codified on a piece of paper. Everyone is free to exercise their will when and where they want. Initiatives percolate up from individuals when and where necessary. Thus, authority can rise (and fall) as freely as ideas and social affiliations. Chaos, disorder, or disregard for the welfare of others need not ensue. Checks and balances can also arise spontaneously.

Authorities tend to rally in response to vicious attacks on their power structures, but might they be lulled to sleep? Couldn't anarchy arise without overt violence or chaos? Might an educated and well connected populace, one that is willing and able to think and speak for themselves in real time, have unwittingly created anarchy? There is good evidence that in many areas of our society anarchy or something approaching anarchy exists today. Consider the bottom-up forces at work in our society today: Artists are free to establish their own styles and change them as many times as they want. Athletes can rise to superstar status overnight. Today's political leaders watch and respond to daily public opinion polls as if their livelihood depends upon them. Business leaders track and respond to consumer buying trends as if their next meal depends upon them. Ballot initiatives place key decisions directly in the hands of voters. There may be something pulling the strings here, but this God works in mysterious ways, for no one can come close to codifying this authority.

At the forefront of science and technology, at the bleeding edge where creativity flows and innovations bubble forth, there are no preestablished rules. Politicians have learned to resist meddling in research plans and university administrators make a point of keeping their hands out of their professors' activities. On the edge of a new frontier where money and hype run wild, trial balloons (from novel sequencing strategies to human cloning) can be floated, support can be sought from a myriad of different sources, and outrage (or indifference) can be openly expressed. Harmful actions can be halted, while neutral or beneficial ones may be carried forward. This new breed of anarchy relegates political leaders, science officials, and business titans to the roles of spokespersons or cheerleaders. It prevents a tyranny of the majority, of rulers, of scientific elites, of monopolistic businesses, and of paternalistic governments. Under anarchy every person is his or her own football, to rephrase a proclamation from an early twentieth-century art movement.[20]

In summation, the race to sequence and characterize all human genes has been a free-for-all. It has been propelled by both selfish and altruistic interests emanating from individuals with a variety of different affiliations.[21] At varying times these interests have overlapped, intertwined, competed, and cooperated, but in totality they appear to be succeeding. Human genes are rapidly becoming known and virtually all of them will become characterized in coming years. Companies such as Incyte and HGS are discovering new gene information and delivering it to pharmaceutical companies in a form that facilitates its use in drug development; the public information base, with the ongoing contributions of HGP and others, is providing invaluable support to the drug development process; the patent process is being used to secure time-limited monopolies, which appears to be motivating the quest for new discoveries and discovery tools; and novel ideas are being transformed into start-up companies which stand a good chance of bringing new products and greater competition to the drug market.*

Perhaps most important, this free-for-all has somehow increased our collective scientific knowledge. Despite corporate efforts to patent genes and impose access charges, the public information base is exploding. The HGP and HUGO (the Human Genome Organization) are organizing a monumental contribution to this public information base; numerous foundations and governments are enthusiastically supporting this effort with a cooperative spirit; new ideas and technologies from the private sector are also helping to propel it; and research findings from the private sector are contributing, too, though not as often or as quickly as many desire. Furthermore, this maelstrom of activity is fostering a revolution in biology and medicine that is likely to raise humankind to a new plane of self-awareness and new control over our physical being.[22]

Where else has such anarchy been witnessed?

---

*The patent process is far from perfect. Overlapping patent claims, unfair claim interpretations, and ambiguous patent rulings often lead to costly and unnecessary squabbles. This incentive system is in need of fine-tuning and oversight, but the current flourish of innovations and new products suggests that it is not broken.

The madcap free-for-all that is the race for the entire sequence of the human genome is a fitting reflection of the genome itself. Over the course of the human genome's evolution, it appears as if a slew of forces—some termed selfish, others as competitive, and still others as cooperative—have fought over, expanded, and otherwise left a mark on chromosomal real estate. Numerous innovations, DNA recombination schemes, mobile sequences, and the like have dramatically altered the course of genome evolution. Not only has the genome been in flux, but it also appears to have adhered to no established rules. Many times biologists have thought that they have uncovered a law of nature—and on an equal number of occasions new discoveries have shown that the "law" has been broken. For example, As don't always align with Ts during DNA replication, proteins aren't always initiated by an AUG start codon, and the Y-shaped chromosome doesn't always bring about maleness. Could our creator be an anarchist?

A taxonomic and historical look at the human genome helps introduce several additional components of biological information that are part of the current revolution in biology. Thus, a brief consideration of genomic relationships and human genome history follows. Onward!

10

Human Genome History

In Gabriel Garcia Marquez's popular novel *One Hundred Years of Solitude* the character Aureliano had been compulsively studying old, cryptically written parchments for years without knowing why he was doing so. Then in the last paragraph of the story he realized that the wondrous history of himself and his entire extended family was written in the parchments, and all at once he "began to decipher the instant that he was living, deciphering it as he lived it, prophesying himself in the act of deciphering the last page of the parchments, as if he were looking into a speaking mirror."[1] The white viscous material derived from our cells is somewhat like these parchments. We may read in it a history of ourselves and of our families. It also speaks to the history of our species and may even yield clues as to why only humans seem to bother trying to decipher it.

There are an estimated ten million living species on earth (most of them microscopic). The genomes of only a couple dozen of them (mostly bacteria) have been fully sequenced, while a few hundred others have been sampled, randomly or otherwise. Yet from this relatively small data set a fascinating account of gene and genome history can be inferred. Not only are all living creatures linked through the use of a common genetic information carrier, but the sequence information itself relates each organism to all others. Thus, DNA from a slug, a bacterium, a redwood tree, or any other species shares sequences so similar to human sequences that it is highly probable that the sequences originated from a common ancestral template. In fact, an evolutionary tree can be constructed solely on the basis of sequence similarities. The picture of evolution that emerges from such analyses uncannily mirrors much of the evolutionary trees that developed prior to knowledge of DNA.

This survey of human genome relations begins with perhaps our most distant cousins, the bacteria. Bacterial genomes typically consist of a single copy of a few thousand genes densely packed on a circular chromosome (about 1 gene per 1000 base pairs). This contrasts with the genomes of most other forms of life—plants, fungi, protozoa, and animals—where genes are arranged on linear chromosomes encapsulated within a membrane (the cell nucleus). Genes from these branches of the tree of life are often interrupted by sequences that do not encode proteins (introns) and are frequently found in duplicate, one from each parent. Among nonvertebrate animals, hybridization, kinetic, and sequencing studies have shown that a total of between 6000 and 25,000 different genes are expressed, whereas among most vertebrates (animals with a backbone, including humans) there are thought to be between 50,000 and 100,000 distinct genes. Vertebrate chromosomes also appear to be more highly structured; their DNA is more intricately wrapped and coiled than those of other groups of species.

The total number of human genes appears to be within the range of other vertebrates, and the number of human chromosome pairs (23) also is within the range of other vertebrates (a Munjak deer, for example, has only seven, while some salamander species have 33). The total human genomic DNA, about three billion base pairs, is also altogether middling (a puffer fish has about 400 million base pairs, and one type of salamander carries about 80 billion base pairs). The human genome thus is like the sun, undoubtedly unique among all stars, but extraordinary only in its proximity to us. Perhaps a closer look will reveal outstanding differences. Could the nucleotide sequence of human genes yield clues to humankind's extraordinary intellect and domination over other species? So far no extraordinary gene features have been discerned.

Genes that perform basic cell functions, such as those involved in protein production and cellular metabolism, so-called housekeeping genes, are typically conserved across all species. Hundreds of bacterial genes have sequences that are similar to those of human genes and perform similar biochemical functions. Some genes from different species are so closely related that one can be substituted for the other. For example, although yeast requires an

RAS gene for survival, substitution of a related human gene results in yeast that appears perfectly normal.[2] For most human genes a similar gene sequence can be found in the genomes of the fruit fly, *Drosophila melanogaster*, the roundworm *Caenorhabditis elegans*, the flowering plant *Arabidopsis thaliana*, or the yeast *Saccharomyces cerevisiae*, four highly studied organisms. Nearly all human genes have corresponding genes in other vertebrate species. Thus, when the leptin gene was found in mice, there was a high degree of confidence in the existence of a human version of the gene.

Cross-species gene searches are routine in molecular biology. A gene discovered in one species can quickly lead to a family of related genes from other species. One can either conduct an electronic search of a database of sequences of a particular species or one can search in the lab by hybridizing (base-pairing) a particular gene sequence (made single-stranded) to DNA from a species of interest (also made single-stranded). The two sequences need not be 100 percent alike; so long as significant portions of the sequences match they will be identifiable by computer search and they will hybridize in the lab. Thus, both in the laboratory and at a computer there are threshold levels of sequence complementarity that enable one to identify related genes. Such genes, called *homologs*, are usually functionally related and therefore provide new avenues of research. Leptin was shown to be involved in fat metabolism in mice; therefore the human homolog of the mouse leptin gene was a great candidate gene for conducting fat metabolism studies in human cells and for finding genetic variations linked to heritable fat disorders.

As more gene sequences became known, a picture of gene sequence relationships across species began to emerge. It is a vast and multitiered web of relationships. Not only are human gene sequences almost certain to match gene sequences within other vertebrates; they are also likely to have similarity with other human sequences. Further complicating matters is the fact that a one-to-one relationship between related genes of different species is often lacking. For example, a set of three similar human genes may closely match a set of four mouse genes, four rat genes, and three chimp genes. These 14 genes may have a low level of similarity to a set of 50 other genes from the same four species.

Despite this problem and the fact that the majority of genes had yet to be fully characterized, a number of researchers have been able to broadly investigate the degree of sequence similarity across species. In one study over 2000 sequences from mice and humans were aligned (entries were found in GenBank and Swiss-Prot databases). The degree of sequence identity in the protein coding regions of these sequences ranged from 36 percent all the way up to 100 percent, with an average of 85 percent.[3] Amino acid sequence identity was also found to average 85 percent. The sequence similarity among related genes varies, since the rate of mutational change varies from gene to gene. Since these sequences were available in part because a high degree of identity allowed them to be found, the 85 percent figure may not be an accurate approximation of all gene sequence relationships between human and mouse. The oft-cited 98 or 99 percent average correspondence between chimpanzee and human gene sequences is based on a far smaller set of genes than the mouse-human study and does not include untranslated sequences. Nonetheless, these studies do portray how extensive sequence conservation can be among genes of different species (see Fig. 10.1).

The differences between monkeys, humans, and rodents are thought to be mostly due to the cumulative effects of slight variations in sequences, particularly in regulatory sequences that determine when and where genes are expressed. One can imagine that by slight alterations in a small number of genes a person could become as hairy as an ape, or an ape could become as smooth-skinned

```
Human insulin:  FVNQHLCGSHLVEALYLVCGERGFFYTPKTRREAEDLQVGQ
Pig insulin:    FVNQHLCGSHLVEALYLVCGERGFFYTPKARREAENPQAGA


VELGGGPGAGSLQPLALEGSLQKRGIVEQCCTSICSLYQLENYCN
VELGGGL--GGLQALALEGPPQKRGIVEQCCTSICSLYQLENYCN
```

FIGURE 10.1   Insulin protein sequence comparison: Human (*Homo sapiens*) on top, aligned with pig (*Sus scrofa*) on the bottom. Each letter represents one of the 20 different amino acids.

as a human. Indeed, there is evidence that supports the concept of atavistic gene variants, sequences that revert at least some aspect of an organism to an ancestral state. A single human gene variant is thought to be responsible for congenital generalized hypertrichosis, a familial condition characterized by excessive facial and body hair, and there is a genetic basis for several developmental disorders that yield more than two nipples.[4] But what about the genetic features that really set humans apart from all other beasts—our incomparable ability to learn, think, and communicate? As of yet no specific genetic features have been shown to confer these traits, and by a variety of metrics the human genome fails to distinguish itself from the genomes from other species. However, the era of genome analysis has only just begun, and many scientists anticipate that genomic correlates to humankind's most remarkable and distinctive attributes will soon be found.

What else might be inferred from cross-species sequence analyses? If one believes that current species evolved from common ancestors, then one must conclude that genes have not only changed over time, but that new ones have been born as well. (A rather unlikely scenario where genes are differentially lost from a mythlike mother of all species could also be imagined.) Where do new genes come from? An organism may acquire genes from another organism. Evidence of such horizontal transfer had been observed, although it is probably rare in animal species where germ cells (sperm and eggs and their precursors) are protected from viral assaults and other agents that may introduce foreign DNA into the germ line. Instead, most genes are created from ancestral genes that are transmitted vertically from generation to generation. New genes (loosely defined here as sequences that convey new information) arise through alterations in old genes. Here the inventive step is in the innovative application of randomly acting forces. These forces may be derived from chemicals, radiation, UV light, or enzymes, agents whose actions are affected by random molecular movements, the alignment of the stars, and the like. Although organisms are remarkably diligent in their protection of the family jewels (they employ hundreds of distinct, finely tuned molecules that act in concert to replicate and maintain DNA sequences), mutations nonetheless happen. They occur, on average, once every 10^9 nucleotide replications, although in certain regions of the genome the mutation rate has been found to be over 1000 times higher. Nucleotides may be altered, deleted from, or added to an original sequence. Resulting sequences may be neutral, that is, they may have no discernable affect on the organism, or they may confer new information, data that could alter the fate of an organism (and in theory could ultimately change the fate of a species or an entire ecosystem). By altering DNA sequences, random mutations can give rise to new gene alleles, which over generations may spread through a population and replace previous alleles.[5] In the meantime populations may split and new species may emerge.

Another route to gene genesis occurs via duplication of a gene or gene fragment. Genes exist in two copies (one on each chromosome) in all but the germ cells, but in the present context *duplication* refers to the inclusion of additional copies of a stretch of DNA. A number of well-understood mechanisms explain the phenomena. During the process of cell division, a portion of DNA sequence may be replicated more than once. The excess copies of the sequence are shed, but may subsequently recombine with the genome. Sometimes chromosomes exchange unequal amounts of DNA in germ cells, leaving one cell with an excess; or viruslike agents may replicate stretches of DNA and reintegrate them into the genome. The consequence of these actions, which act somewhat randomly, may be a composite of two or more gene fragments or a duplication of an entire gene.

Duplicated genes may then diverge in sequence through the types of mutations just described. Evidence for such events exists in DNA sequences. For example, there are two opsin genes located adjacent to each other in most humans. In the retina, opsin protein transduces light energy into chemical energy, thereby enabling sight. The adjacent opsin gene sequences differ in only 2 percent of the nucleotides, but whereas one opsin protein is most sensitive to red light, the other is most sensitive to green light. Further molecular taxonomy suggests that early in the course of vertebrate evolution the opsin gene duplicated and diverged. Presumably, by gaining an additional opsin gene our animal ancestors could better distinguish colors and thus better survive.[6] Each new gene is also another stepping stone for the emergence of additional genes. Duplication and divergence results in gene families, sets of genes (both within a species and across different species) that have sequence similarity.

Over a hundred years ago, Charles Darwin proposed the theory of natural selection to account for evolutionary change. This theory asserts that particular heritable traits, ones that provide a reproductive advantage, will spread through the population and thereby change the entire species. There is little doubt that human traits and the genes that enable them have been shaped by such forces; however, a simplified interpretation of natural selection provides a poor account of the dynamics of genetic changes and at best a hazy explanation for the similarities and differences between DNA sequences of humans and other species. The rate at which a new gene, beneficial or otherwise, spreads through a population depends on a variety of complex, interconnected factors, including DNA recombination, population dynamics, and environmental conditions.

Recombination is particularly noteworthy here. Within all species there are mechanisms for exchanging DNA sequences among members of the species. Nonbacterial species often utilize sexual recombination, an inventive means of shuffling parental gene sets. Indeed, sexual recombination is a tremendous innovation, for not only does it provide a wellspring of new genes by making duplicate genes through unequal cross-overs and by cutting out and pasting together different gene fragments, but it also creates innumerable novel gene combinations.

Sperm and egg join to initiate development of new life. These germ cells each have a single set of 23 chromosomes, which they receive when they are formed through the division of primary spermatocytes and primary oocytes. Each of these germ cell precursors has two sets of 23 chromosomes, one from each parent just like the rest of the cells of the body. Within the primary spermatocytes and oocytes, paired chromosomes break and rejoin in numerous places so that offspring inherit a random mix of the genes of their four grandparents. The gene combinations are innumerable and always changing, since new genes may be introduced at any time. Reproductive success may be influenced by the combinatorial effects of a new gene, a second allele, and other genes, as well as by environmental conditions, which also tend to be complex and varying. Therefore, interpreting cross-species gene sequence differences is not simple. It may be much easier to explain human bipedalism and finger dexterity in terms of natural selection than it is to explain human genes that differ from primate genes in a few dozen nucleotides.

Natural selection harvests what bubbles forth and with god-like power shapes a population. But if natural selection were a deity it would share its realm with another authority, the god of chance. Chance encounters may initiate relationships that lead to children. Chance events can bring about untimely deaths. And at the molecular level, chance affects the occurrence of mutations and their distribution in the population, particularly those that have little or no discernible effect and are thus immune to selective pressures. Such mutations occur in the vast tracks of DNA that do not encode expressed genes or in regions of expressed genes that do not affect traits. Since there is redundancy in the genetic code (most amino acids are encoded by more than one nucleotide triplet), mutations may alter a codon without altering the encoded amino acid. Even changes that alter the amino acid sequence of a protein may have no detectable affects on an organism, as some amino acids appear to act merely as space-holders. All of these neutral changes can manifest themselves in the genome and further complicate analyses of cross-species sequence comparisons.

Genome mapping and sequencing studies have revealed that some more sporadic types of recombination have also helped to shape the human genome. Since the number of different chromosomes differs from species to species, large genomic rearrangements must have occurred as current species diverged from ancestral ones. As with other types of mutations, evidence for such events can be readily observed across a single living generation. People are born with additional chromosomes (*trisomy*), chromosomes that contain a fragment from another chromosome (*translocations*), or with a fragment that is flipped in orientation relative to the rest of the chromosome (an *inversion*). These events often lead to disability, sterility, or death, but some of them can also readily spread through a population. Detailed mapping studies have shown that the 20 chromosome pairs of the common mouse, *mus musculus*, and the 23 pairs of human chromosomes can each be cut up into over 100 fragments that almost precisely match each other in terms of gene composition and order. This suggests that a measurable degree of large-scale restructuring has occurred in the course of mammalian genome history.

Vertebrate species tend to have four times as many genes as invertebrate species, and it is often possible to match a set of four vertebrate genes with sequence similarity to a single invertebrate gene. The quantum leap in total gene number in the vertebrate lineage is believed to have been caused by two ancient genome doubling events, each followed by large-scale gene divergences. The first doubling is believed to have occurred in an ancestor of all vertebrates living about 550 million years ago. Like countless gene mutations and genomic rearrangements, a genome duplication could have occurred randomly, either by the joining of two embryonic cells or by DNA replication without cell division. Cells are somewhat accustomed to having twice as many copies of their genome, for they must hold a double load for a period of time prior to dividing.

An organism that survived with a double genome would pass two copies of 20,000 or so genes to its offspring. Like all genes these duplicates would then face the prospect of extinction through mutations that reduce them to junk DNA status. However, they would also be free to mutate into genes that increase the organism's chance of survival. Large tracts of non-protein-coding DNA might have helped out. Such DNA, in and around genes, may act as a large target for mutations that could alter gene regulation. If the duplicate genes were expressed at distinct times and locations, then they could serve two distinct functions and would thus survive. Alternatively, duplicate genes could differentiate themselves through mutations in the less expansive coding regions. There is abundant evidence within gene families of both these events. Members of the kallikrein gene family, for example, have similar sequences and encode proteins with similar enzymatic activity, but some of them are expressed in vastly different sets of tissues and thus have vastly different functions.[7] Pancreatic/renal kallikrein is involved in regulating blood flow, while Kallikrein 3 (also known as Prostate Specific Antigen, or PSA) is made in the prostate, where it is believed to regulate cell growth. On the other hand, members of the bcl-2 gene family may be expressed in the same cell types, but carry out distinct functions due to variations in their respective amino acid sequences.[8] Bcl-2 protects against programmed cell death (*apoptosis*), while its sister protein BAX promotes these cellular suicides.

Another genome doubling is speculated to have occurred about 50 million years after the first one. Lineages that retained the two-fold increase in genomic content, a total of about 40,000 genes, include the hagfish and a few related species of jawless fish. Other vertebrate lineages grappled with the four-fold increase, 80,000 or so genes, which resulted from this additional doubling of the genome.

As has been noted, at any given time a vertebrate cell will express only a fraction of its total number of genes. About 60,000 genes are silenced in any given cell type. The development of a large-scale gene silencing mechanism may have facilitated the quadrupling of the genome by enabling the differential expression of recently duplicated genes. This in turn may have enabled the evolution of more complex development pathways and body structures. Adrian Bird and Susan Tweedie of the University of Edinburgh have suggested that DNA methylation, a reversible modification of nucleotides that has been shown to reduce gene expression, occurred simultaneously with the emergence of a fourfold greater number of genes in vertebrate species.[9] DNA methylation is considered to be an epigenetic change, a heritable alteration that occurs external to the sequence of nucleotides. Perhaps there was a comparable innovation that allowed humans to distinguish themselves from all other species. Or perhaps humans are destined to create such an innovation.

The forces that shaped our genomes are of course still operating today. Mutation, recombination, the birth and death of genes, selective pressures, and randomly acting forces all continue within *Homo sapiens*. When one aligns DNA sequences from any two individuals, on average about one in a thousand base pairs will differ. The nature of these differences and their significance will be considered in the next chapter.

11

Comparing Human Genomes

Just whose genes are being sequenced? This question surfaces repeatedly and is a jumping-off point for a myriad of questions and concerns about large-scale human gene sequencing. The answer is that, at least initially, both cDNA sequencing and genomic sequencing projects utilize DNA derived from a variety of individuals.

The pathway to any human sequence begins with living tissue. DNA or RNA is extracted from surgical waste, such as a tumor tissue or tonsils, organ tissue from recently deceased individuals, blood samples, or other bodily material. A variety of volunteers donate a portion of their flesh. Thus, somewhat like Mary Shelley's Dr. Frankenstein creating his monster, the databases piece together a mosaic of human sequences in the hopes of forming (in code only) a composite sketch of the whole genome or at least all of the genes. The EST factories and the Human Genome Project used tissues and cells from thousands of different individuals. Celera scientists, in contrast, determined that it was preferable to apply their shotgun sequencing technique on only a small number of samples. They eventually settled on five individuals, three females and two males, who had identified themselves as Hispanic, Asian, Caucasian, or African-American.[1]

Generally, sequences cannot be traced to particular donor names, as this information is usually made to be irretrievable. Privacy and discrimination issues are still of concern and are likely to become even more important as the databases become more complete and as advances in technology make it easier to sample, sequence, and analyze anyone's DNA. These concerns present additional reasons for people everywhere to educate themselves as to what's going on with human genetics research.

Human sequence data, whether a mosaic or otherwise, acts as a reference to which sequences from any individual can be compared. Such comparisons have been a top objective from the start of the Human Genome Project. Francisco Ayala stated at the 1986 human genome conference in Santa Fe:

> It is usually said that the human genome consists of about three billion base pairs. This is an understatement by ten orders of magnitude. It would be more accurate to say that the human genome consists of ten billion times three billion base pairs. All humans have two haploid genome complements, each with three billion base pairs . . . and there are five billion humans around . . . if we want to understand humankind and ourselves, it will not be enough to have one sequence. We have to understand the variations.[2]

Indeed, genetics has always been about variation. By observing or manipulating discrete variations in molecules, cells, tissues, and physical or behavioral characteristics, the relationships between these variations can be deduced. The white eyes of fruit flies are associated with lack of a particular pigment. A change in nucleotide sequence of a tumor suppressor gene leads to uncontrolled cell growth in cell cultures. Such causal links provides a basis for identifying gene functions, as well as for developing therapeutic, diagnostic, and prognostic targets and leads. The sum of such causal links provides a framework for understanding living organisms.

Genetic variations play a part in a wide spectrum of human characteristics, including differences in physical traits and at least some thought processes and behaviors. Great strides are now being made in delineating the sequence of all human genes, including gene variants, and in elucidating all gene functions. Monstrous databases are coming into being and the goals of the sequencing projects are being realized. The enormity of the implications of these feats is not lost here.

Medical and academic interests in the similarities and differences among people pale in comparison to a deeper and more universal interest. It is an interest that affects virtually all human interactions. An ant is just an ant, but a person is never just a person. We take notice, consciously or not, of how we look and act. Hints of one's genetic heritage are present in a person's birthplace, ethnicity, religion; in the size and shape of one's body; in the color of one's hair, eyes, or skin; in one's visible defects and disabilities; in the characteristics of one's family members and ancestors; and in countless other ways. People of all cultures analyze and act on such information. Nongenetic factors, such as child rearing, education, experiences, culture, and customs play enormous roles in human individuality, but genetic factors can never be ignored. The interplay between the two is complex, mysterious, and controversial, and when genetic findings reach our collective consciousness it is like a submarine surfacing in an embattled sea.

The genome is integral to every cell of the body. Gene therapy notwithstanding, any one person's genome is rather rigidly defined. In some ways it can be considered a hard-wired code of limitations and restrictions, of disease propensities and behavioral proclivities. A good portion of one's life is spent in testing and exploring one's physical and mental capabilities, in exploring one's own operating system. There is concern that people may be robbed of the wonders of life's mysteries and the benefits of self-discovery if the human genome is understood. There is also concern that a person's privacy or freedom may be robbed by those who read or interpret his or her DNA sequences. There is anxiety over the possible assertions of behavior genetics, for an estimation of a person's capabilities provides a basis for determining the levels of responsibility that we are held to. If those capabilities are genetically determined, a number of difficult questions follow. What criminal behavior should be pardonable? What treatment does a drug addict, compulsive gambler, or smoker deserve?

Genetic assumptions also provide a basis for determining group membership, with its associated battles. They may erupt in genocide, flare up in genetic discrimination disputes, or simmer subliminally in everyday human interactions. Though they are driven primarily by compelling medical applications, genetic information and enabling technology as well as the people that produce them are brought into the fray. Thus, although the nature/nurture debate is not the topic of this book, it seems that a perspective on our current knowledge of the genomic diversity is appropriate.

When viewed from atop a tall building, people on a street below do appear like ants. Most distinguishing features are indiscernible. So, too, are human genomes when viewed through an ordinary microscope. They appear as 23 sets of chromosomes, beautifully stained to reveal distinct sizes, shapes, and stripes, but indistinguishable from most others. The karyotype (display of chromosomes) reveals the sex of the individual. It may also reveal a particular narrowing of the X chromosome that is associated with a particular type of mental retardation (fragile X syndrome), a third copy of chromosome number 21 (which causes Down syndrome), or the unusually long chromosome of members of Roger Donahue's family (from Chapter 3), but a karyotype does little else to distinguish people. It is at the sequence level that all human genomes are found to differ. Across any two sets of human chromosomes gene order and spacing usually agree, but the sequences themselves are riddled with variations. If the genomes of any two individuals were aligned, then between 0.1 and 0.2 percent of all the nucleotides would not match. Alternatively, one can say that 99.8 to 99.9 percent of the sequences precisely match. Within the protein-coding sequence of genes there is slightly more sequence identity.

Each person has a unique set of polymorphisms (sequence variations). Hence, it is possible for a DNA-based test to distinguish any one person from all others. If one tried hard enough, one could even find variations between the DNA of identical twins. From the other perspective, we each have unique bodies that are, in part, a manifestation of our unique sets of polymorphisms. Since populations exist that have distinct heritable characteristics, such as skin color or facial features, and have lived apart from others for many generations, it seems safe to assume that members of these groups share distinct sets of polymorphisms. In addition, distinct populations or ethnic groups have distinct incidence levels for genetically determined traits and diseases. The gene variant that enables adults to digest the milk sugar lactose swept through Europe with the introduction of dairy farming thousands of years ago, and today adults of Norwegian descent, for instance, are much more likely to have this gene variant than those of Japanese descent.

Nonetheless, despite such observable genetic differences, several different studies have shown that there is more genetic variation between individuals within a population than between populations.[3] A detailed study of 16 distinct populations (native Mbuti pygmies from Zaire, Cambodians, Japanese, Northern Italians, Chinese-born San Franciscans, etc.) showed that differences between members of the same population accounted for 84 percent of all variability, whereas only 16 percent was due to differences between populations. Such experiments discredit the notion of human races, the concept of genetically homogenous (or pure) groups of people! Although there are sets of gene variants that determine commonly held skin color[4] and other features, members of such populations have a far greater number of variants not commonly held, for example variants in genes that influence such things as height, weight, blood pressure, bladder size, particular disease-fighting abilities, and so forth. This implies that a person of Jewish heritage and another from Palestinian heritage, for example, will often be more genetically alike then a pair from either group.

Perhaps the most striking genetically determined variation is that between males and females. Remarkably, a single gene, called the *testis determining factor* (TDF), may turn a fertilized egg from a pathway of female development to one of male development. TDF is present in males (on the Y chromosome) and absent in females. It is the gene that resides at the SRY locus that was positionally cloned in 1990 (as was described in Chapter 3). Many other genes are subsequently involved in the process of sexual development, which usually, though not always, leads to one of two physiologically distinct types. These genes, ones that produce sex hormones, for example, exist in both sexes. It is their differential expression that is responsible for the characteristic differences between males and females. Nonetheless, the dichotomy is rooted in the presence or absence of a particular gene.

Total gene number may vary in additional ways. Consider the following two scenarios: (1) The birth of a gene. A single gene duplication event occurs in a single individual. So long as the duplicated gene is not harmful, it may slowly, over many generations, move through the population and at some point diverge (through the accumulation of mutations), creating a new gene. Under the influence of natural selection and chance this new gene may slowly spread through the population. At any given moment some individuals will have the gene, while others will not. (2) The death of a gene. A gene can be rendered inconsequential due to changes in the environment or through the introduction of new genes. Mutations could then accumulate that render the gene nonexpressed or otherwise inoperable. In the course of many generations the gene could gradually become unrecognizable, like a carcass becoming absorbed into the soil. Dead and dying genes may litter the genome. They may show some resemblance to living genes, revealing bits of sequence similarity, yet be completely nonfunctional.

Consider the L-gulono-gamma-lactone oxidase gene. In rats and other nonprimate animal species, L-gulono-gamma-lactone oxidase catalyzes the last step in the synthesis of vitamin C. Human bodies, however, do not make the popular antioxidant. Scientists speculate that 25 million years ago our ancestors did make it. In 1994 researchers in Gifu, Japan, exhumed a sequence within modern humans that resembles the rat L-gulono-gamma-lactone oxidase gene.[5] The human gene remnant appears to have had numerous nucleotide substitutions, deletions, and insertions. Exons 8 and 9 are completely missing, and stop codons interrupt the remaining reading frames, but the sequence is still similar enough to the rat gene to imply that the two must have had a common ancestor. Our forebears who lost the gene were able to survive because its disappearance coincided with a change in their diet, the addition of vitamin-C-bearing fruit, but for many generations those with the gene must have coexisted with those lacking it. Such gene variation is also certain to occur in today's human population.

The total human population has risen meteorically from about 100,000 to about 6 billion over the last 5000 generations. Environmental changes, including human-induced modifications in sanitation, health care, and nutrition, have neutralized the impact of many sequence variants that would otherwise have been subject to negative selection pressures. Meanwhile, the forces that produce mutations have continued to act on human DNA. Neutral mutations are less likely to disappear in an exponentially expanding population than in one that contracts or stays the same. Therefore, the total number of distinct genes and gene variants within the entire human species has probably grown considerably. Human languages and traditional cultures may be falling by the wayside, but both old and new DNA sequences are being retained.

Generally, a diverse set of variants, a large gene pool, provides a survival advantage for a species. A rare bacterium with a gene that enables it to survive an antibiotic assault can help revive an otherwise-doomed bacteria population. The human population also has rare sequence variants, which under certain conditions can make the difference between life and death. An example of this is the rare CCR5-delta-32 gene variant. The CCR5 gene encodes a protein that sits at the surface of certain immune system cells, where it helps relay important messages in the coordinated battles against diseases. HIV uses CCR5 to gain entry into these cells. In 1996 researchers found that the rare delta-32 variant, with 32 less base pairs than other alleles, confers a strong resistance to HIV infection.[6] Thus, CCR5-delta-32 has as much protective power as a bulletproof vest. The prospect of more such sequence variants and their potential utility in developing new treatments has not escaped notice. There are likely to be genes and gene variants that are of little or no significance until the local environment changes and individuals are challenged with disease, starvation, or other hardships. Since there are countless different types of environmental insults, it is rather fortunate that there also are innumerable distinct genes, gene variants, and gene combinations.

The composite human genome, one that encompasses all human genomic diversity, has far more than three billion base pairs of sequence and far greater than 100,000 different genes. Large-scale sequencing projects are likely to uncover many rare genes as well as the remnants of ancient genes. An understanding of such sequences is likely to facilitate and hasten the deployment of new disease detection and treatment options. Scientific enterprises thereby enjoin, enhance, and accelerate other ongoing natural processes that generate, preserve, and employ novel genes and gene combinations.

Scientific work brings certain truths to our awareness, to our collective consciousness. Conscious processes are tempted, lured, enticed, cajoled, urged, encouraged, or forced into usurping unconscious ones. Nowadays, not only do scientists compute the fiber content of your breakfast cereal, the sugar and fat content of your snacks, and the vitamin content of the fruit you eat; there is also a fair chance that you actually use this information in choosing what to ingest. With every meal you may integrate a slew of internal urges, advertising images, nutritional and health data, and the interpretation of such data by your mother, your nephew, and your hairdresser, the id, ego, and superego. Genetically programmed noncognitive mechanisms for securing an appropriate blend of nutrients are being bypassed and manipulated by our conscious minds and/or by a collective consciousness, the scientific knowledge that settles out from all the research findings.

Our genetically-rooted inclination to size each other up (it occurs among individuals of almost all species) is poised to leap further into the conscious realm. What people will do with these new powers, and these new burdens, remains to be seen. Clearly, there is the potential for both benefit and harm, but before entering the battlefield it may be worth considering certain natural limitations in our scientific powers. The degree to which genomic sequences may be predictive of physical or mental traits may be less than many imagine. These limitations and the scientific revolution they are inciting is the subject of the following chapters.

12

A Paradigm in Peril

Genetic disease genes are being picked off like ducks in a shooting gallery. Genes that cause major diseases, such as cystic fibrosis, breast cancer, colon cancer, and Alzheimer's disease have been conclusively identified. Sequence variations responsible for hundreds of rare or less well-known conditions, such as congenital myasthenic syndrome, Alport syndrome, Smith-lemli-opitz syndrome, mitochondrial encephalomyopathy, Niemann-Pick disease, dominant cone-rod dystrophy, and hemochromatosis have also been pegged. Many more human diseases and traits have been linked to specific chromosome regions. These stretches of DNA reportedly make people vulnerable to obesity, alcoholism, depression, violence, homosexuality, or numerous other conditions. Presumably, other alleles help people to be fat-free, sober, happy, peaceful, and heterosexual.

"Gene for Intelligence Found!" proclaims a newspaper headline. "Gene for Baldness Discovered!!!" screams another. Bravo to the many journalists who sensationalize genetic findings! They alarm only the alarmists, con only the gullible, and taunt only the tauntable. In others they may induce skepticism and/or curiosity. What is really going on here? The pace at which such reports are being made appears to be accelerating. Is there anything that is not heritable, and will each heritable trait be fully explained by the action of genes, as a puppet's motions are explained by the movement of a set of strings?

No—in fact the entire framework of shared assumptions that has led to such conclusions is in peril. Genetic determinism, the dominant model of biology theory since the rediscovery of Mendel's work a century ago; the model that places "the gene" at the center of the universe, that considers single genes as master regulators and hierarchical leaders directing virtually all human traits and diseases, that reduces macromolecular phenomena to discrete, linear molecular events—this paradigm of the life sciences—is in its dying days. The current flourish of definitive proofs of trait-causing genes that appear in scientific journals is a swan song. The slew of indefinite reports that allude to, but do not prove, causal links between genes and traits, and the distorted interpretations these reports generate are the cackles of a dying beast.

A beautiful model of life, a profound explanation of biological phenomena, the research framework upon which deadly disorders have been dissected and reduced to singular dysfunctional molecules, a framework validated and solidified by lifesaving new treatments for diseases such as diabetes and hemophilia, a robust predictive science of disease; all of this is winding down. Of course, insulin will still be used to treat diabetics; the addition of factor VIII clotting agent will continue to protect the blood of hemophilia patients; and the drug development and diagnostics pipelines will remain filled with the legacies of Mendel's pea work for years to come. The gene deterministic model is not wrong. It is simply not enough. It has been stretched beyond its capacity to explain biological phenomena. It is being exhausted of explanations. It is being confounded by more and more unexplained phenomena. You don't need to be Albert Einstein to see the cracks in this paradigm.

The degree to which biological phenomena have been explained through gene discoveries is not nearly as extensive as the flood of reports suggests. First, many of the reports are of statistical associations between polymorphisms and traits. The polymorphisms serve as markers that define regions of a chromosome spanning up to a hundred million base pairs. Presumably, a specific trait-causing gene variant lies within these regions. Oftentimes, however, additional data discredits the original statistical claims. Other times extensive searching leaves exhausted researchers empty-handed and befuddled. They may then speculate that there is not one gene culprit but a "bad neighborhood" that hides several thugs whose combinatorial actions or non-Mendelian modes of inheritance foil the gene-seekers. Non-Mendelian inheritance patterns may be caused by epigenetic phenomena like DNA methylation or by gene sequences that are particularly prone to mutations.

Second, trait or disease gene discoveries often are correlative, but not predictive. Individuals with the gene variant do not necessarily have the trait or disease; instead they merely have an increased likelihood of possessing the phenotype. In genetic terms, the penetrance of the gene variant may be low, and other factors, such as environmental conditions, other gene variants, or random processes, may work in combination to determine the phenotype. Many times such extraneous factors are alluded to, but they are almost never explicitly identified and incorporated into a predictive model that is as good as, for example, the one that explains sickle cell anemia.

Third, identified disease genes are often responsible for only a small fraction of the disease incidents. Common afflictions such as diabetes, Alzheimer's disease, and atherosclerosis have multiple causes, and only a tiny portion of the cases can be attributed to specific gene variants.

Last, the identification of the guilty sequence may do little to explain how the disease manifests itself. Therefore, short of replacing the sequence in every cell of the body, a means of altering the course of the disease is not apparent. "The disease gene is a starting point, the first link in a molecular chain of events, an important early milestone in the long path towards a cure," bleat the flock following each and every disease gene discovery. Such statements are probably valid, but just how will knowledge of each disease-causing gene bring about a better treatment or a cure? Even if all the molecular interactions that emanate from the disease gene are delineated, it is not at all clear under the current framework whether an adequate predictive model—one which would allow treatments to be rationally designed—can be constructed. Consider some of the perplexities of the following two human conditions.

Breast cancer is an often deadly disease affecting one out of every eight woman in the United States. A small percentage of cases occur in comparatively young women from families that have a history of many such tragedies. After an intense search, a gene variant specific to these patients was identified in 1994.[1] Fourteen months later, a variant of a second gene was uncovered from a different set of families.[2] Hopes were raised for both the development of gene-based diagnostic tests and the delineation of the molecular mechanisms by which these genes act. Progress on both fronts came quickly. Hundreds of variants of the two genes (which had been christened BRCA1 and BRCA2) were identified and many of them were found to correlate with a high incidence of early onset breast cancer. Within a few years tests that could distinguish between these variants were on the market, providing doctors and patients with a much better means of assessing breast cancer risk. Women with either a faulty BRCA1 or a faulty BRCA2 have about an 80 percent probability of getting breast cancer by the age of 70.[3] However, family inheritance patterns indicate that only about 5 percent of all breast cancers are caused by heritable factors.[4] Generally, neither the BRCA1 nor the BRCA2 disease variants are found in the nonfamilial or sporadic cases. Furthermore, as much as one-half of the heritable cases are attributable to neither the BRCA1 nor the BRCA2 variants. In these cases a clear Mendelian pattern of inheritance is absent, and there is every indication that a far greater entanglement of environmental and multiple genetic forces are at work.[5] Thus, additional single breast cancer gene variants that will encompass the remainder of breast cancer cases, putative BRCAs 3, 4, and 5, are not on the horizon.

Simultaneous with the diagnostic advances, a battery of tests were conducted to determine the mechanisms by which BRCA1 and BRCA2 control breast cell growth. Both genes appear to act in maintaining the integrity of DNA sequences.[6] The mouse versions of BRCA1 and BRCA2 proteins interact with a protein that is known to function in the repair of DNA damage, and the loss of either of the two induces a characteristic DNA damage response. DNA damage has been shown to result in gene mutations that bring about uncontrolled cell divisions, the hallmark of all cancers. The two-hit cancer hypothesis, in which an inherited (germ line) gene variant greatly increase the chances of this new gene variant (a somatic cell mutation), has great support, particularly in explaining certain colon cancers. The BRCA-induced breast cancers may also fit this model. Perhaps the BRCA gene discoveries will act as levers for prying open a dark box that holds the secrets of how cancer develops, not just in the families in which the deadly BRCA1 and BRCA2 variants are found, but in other patients as well.

Such are the great hopes that accompany all highly touted gene discoveries. However, by all accounts an enormous amount of work remains to be done before the riddles of BRCA-induced breast cancer are adequately solved. When research proceeds one gene at a time (i.e., determining how BRCA1 interacts with protein X, how protein X interacts with protein Y, and so on), then the progress may be painstakingly slow. Furthermore, if the molecular pathway is not linear, but branches in complex ways (as appears to be the case with BRCA-induced breast cancers), then a simple and satisfying understanding may be elusive. At this point, the prospect of a BRCA1- or BRCA2-inspired drug seems remote. Even the introduction of a healthy BRCA gene in every tumor cell may do nothing to stop a breast tumor, for the genetic makeup of the cancer cells has already changed in innumerable ways. Breast cancer, even in the small percentage of cases with a known genetic cause, may warrant a fundamentally different approach than that which worked so well for hemophilia and insulin deficiencies. Undoubtedly, members of the families identified with the disease-causing BRCA variants anxiously await a new approach.

The databases of scientific literature are littered with reports of distinct chromosome regions implicated in the development of schizophrenia, a chronic and disabling mental illness that afflicts nearly 1 percent of the population. For decades it has appeared as if researchers have been on the verge of cracking this genetic puzzle and identifying one or more schizophrenia-causing genes. A preponderance of evidence unambiguously demonstrates that in many people heritable factors contribute to the development of the disease. For example, a compilation of several studies show that monozygotic twins of schizophrenics (who have nearly identical sets of gene variants) have a 48 percent chance of also being affected, whereas dizygotic twins (who on average have half of their gene variants in common) have only a 14 percent chance.[7] Families with a history of schizophrenia provide the raw ore, the DNA samples and pedigrees, from which earnest geneticists seek to mine gene variants responsible for schizophrenia.

However, on this treasure hunt many holes have been dug, but no gold has been found. In some of the cases the link between the implicated chromosomal region and the disease may have been coincidental. In other cases a link may have been concocted out of poorly applied statistics and wishful thinking. But the greater the adversity, the greater the triumph. Every time a balloonist failed in an attempt to circumnavigate the world, the goal appeared more elusive and the prospect for heroism grew. So too with the hunt for a schizophrenia gene, and therefore it is no surprise that additional resources have been secured and the hunt goes on. But even a huge international collaboration, designed to strengthen the statistical analysis by bringing in DNA samples from a greater number of stricken families, has failed to yield any guilty gene variants.[8] Might there be a better way to approach the problem?

The effect of any single schizophrenia-associated sequence variants must be diluted by other genetic and nongenetic factors. The twin studies previously stated imply that nongenetic factors also play a significant role in the development of schizophrenia, for if genes alone were responsible, then close to 100 percent of the monozygotic twins of schizophrenics would be affected. Most participants in the struggle to find schizophrenia-causing genes now admit that variants in a single gene are not likely to account for very many of the total number of cases. Instead, it is highly likely that multiple genes and idiosyncratic environmental factors are involved.[9] Breast cancer and schizophrenia, like many of the most common and most intractable human diseases, are not attributable to single gene variants in the majority of the cases. Rheumatoid arthritis, Alzheimer's disease, multiple sclerosis, depression, addictive behaviors, allergies, asthma, diabetes, cancers, atherosclerosis, obesity, and many other diseases are considered to be complex, polygenetic, or multigenetic disorders. For each only a small fraction of the cases can be attributed to single genes. For the vast majority of patients a combination of several genes and/or environmental factors are believed to underlie the disease process.

Although monogenetic disease triumphs have attracted the greatest amount of attention, the challenge of complex disorders and traits has not been ignored. Complex disorders and traits present tremendous unmet medical needs. Dozens of new companies, as well as established drug companies and noncommercial labs, have been mobilized to confront such complex disorders and traits. However, as of yet it is difficult to find even a single example of a complex trait that has been reasonably explained in terms of a network of identified gene variants and environmental factors. At best, complex disease and traits have been partially distilled to a half-dozen lambda values, which are measures of the increased chance that an individual with a particular variant will contract the disease or trait. Odds such as these are the type of fodder that feeds the interests of insurance companies and gambling houses, but what can individuals and their doctors gain from them? Suppose you do not have cancer but tests indicate that you have a gene variant that is three times more common in colon cancer patients than in others. You may be a bit nervous (perhaps, depending upon whether you have a gene variant for nervousness) and get a test for colon cancer. Your odds of contracting the disease and the costs of diagnosing and treating it could be fed into a computer, and health care policies could be created that would save lives and costs, but is this the best that science has to offer? The mandate of biological sciences is to provide better explanations of life, more accurate predictions of biological phenomena, rational intervention schemes, and more effective medicines. Maintaining "the gene is king" frame of mind does a disservice to these goals.

The names of genes often give an inaccurate picture of gene function and contribute to, or at least reflect, the lofty status individual genes are given. Genes are given names by their discoverers that usually either reflect a proven or presumed molecular function or the phenotype that is a consequence of a particular variant or mutant. Thus, the protein encoded by the L-gulono-gamma-lactone oxidase gene oxidizes the vitamin C precursor, L-gulono-1, 4-lactone. BRCA1 is the first *BReast CAncer* gene, and presenilin-1 (for senility) is an Alzheimer's disease gene. However, as indicated, the occurrence of disease genes is often in concordance with only a small fraction of the cases. They do not hold dominion over the entire realm. The presenilin-1 variants that lead to Alzheimer's disease occur in less than 5 percent of Alzheimer's disease patients. Furthermore, genes often have more than one function. One molecule, known by at least four different names (AMF/NLK/PHI/MF), has been shown to bind to receptors that stimulate directional cell movement, bind to other receptors that promote the survival of certain neurons, catalyze the conversion of glucose-6-phosphate to fructose-6-phosphate, and mediate the terminal differentiation of certain tumor cells.[10]

A satisfying biological understanding of a disease process is one that shines a light on pathways that lead toward better treatment options, or at least suggests a means of prevention. In most cases, the identification and initial characterization of a disease gene does not by itself accomplish this. And, despite the flood of gene discoveries, it seems unlikely that the identification of 100,000 or more genes and gene functions will by itself bring a satisfying understanding of the most common and debilitating human conditions. In trying to put together a puzzle of more than 100,000 pieces, getting all the pieces is certainly good news, but the challenge of fitting them all together remains.

Necessity may be the mother of invention, as the adage states, but there are other generative forces. Authority, the father of rebellion, may play a comparable role. Reductionism is a process by which problems are dissected into smaller and smaller components. The authoritative or dominant positions of reductionism and its cousin, gene determinism, in the biological sciences have been under attack for some time. Richard Lewins and Richard Lewontin wrote in 1985 that, "despite the extraordinary successes of mechanistic reductionist molecular biology, there has been a growing discontent in the last twenty years with simple Cartesian reductionism as the universal way to truth. . . . Holistic, structuralist, hierarchical, and systems theories are all offered as alternative modes of explaining the world, as ways out of the cul-de-sacs into which reductionism led us."[11] Over a decade later, mechanistic and reductionistic biology has grown even stronger and sequencing projects are now providing convenient hooks on which to hang messy, unexplained biological phenomena. Yet, critical voices remain. For instance, in 1997 Richard Strohman, an emeritus biology professor at the University of California at Berkeley, wrote that:

> in promising to penetrate and reveal the secrets of life, [genetic determinism] has extended itself to a level of complexity where, as a paradigm, it has little power and must eventually fail. The failure is located in the mistaken idea that complex behavior may be traced solely to genetic agents and their surrogate proteins without recourse to the properties originating from the complex nonlinear interactions of these agents.[12]

Strohman suggested that a Kuhnian revolution was in the works.

The scientific historian and theorist Thomas Kuhn asserted that in order for one paradigm to fold a new one must overcome it.[13] No matter how inadequate, scientific paradigms do not simply crumble under their own weight. A new contender must offer a more compelling framework. Whether generated out of necessity, rebellion, or creative joy, the new paradigm must outdo gene determinism in explaining difficult biological phenomena and providing pathways for the development of better medicines. There are many contenders, particularly outside of traditional science. But, as they exist today, neither voodoo, homeopathy, food supplements, nor other alternative medicines will do. For each effective herbal remedy, for example, there are probably a thousand ineffective or even harmful ones, and consumers have no adequate means for sorting through it all. The new paradigm must still harbor the principles of science or be so damn compelling as to eclipse them.

How will it be known that a new paradigm has arrived? When that paradigm does what science should do: hits the target with accurate predictions. It should provide solutions that individuals and their doctors can rely on—a diagnostic test for prostate cancer or Alzheimer's disease that provides a satisfying level of accuracy; a concoction that will halt memory loss, epilepsy, or Parkinson's disease; a pill that will send a tumor into oblivion or reverse the course of multiple sclerosis—solutions that are as compelling as the ones developed generations ago for polio and tuberculosis.

An alternative to gene determinism has yet to assert itself. But, as will be described in coming chapters, the technological revolution that enabled large-scale gene sequencing projects is also revolutionizing researchers' approaches to biomedical problems. Perhaps, an Einstein-like individual will lift the life sciences up to another plane, or perhaps the breakthrough will be accomplished by a network of people who are just really smart.

13

The Ancient Internet— Information Flow in Cells

What is the nature of information exchange within cells and how can information theory and information science help in understanding molecular communications? Claude Shannon, the founder of information theory, never gave much consideration to the transmission of information within living organisms, but other information scientists, as well as a few linguists and electrical engineers, have.

One application of information theory is the evaluation of the information content of the human genome. Human sequences have roughly an equal number of As, Ts, Cs, and Gs. Each nucleotide contains 2 bits of information; a gene of 2000 base pairs has 4000 bits (or 500 Bytes, since 8 bits = 1 Byte); and the genome has 6×10^9 base pairs or 1.2×10^{10} bits (1.5×10^9 Bytes). A single compact disk can barely hold the information content of a single human genome. With a volume of about only 6×10^{-19} cubic meters, the nucleus of a human cell has a rather dense information storage capacity. This has led a number of people to consider the use of DNA as an information storage device, and in a few instances to demonstrate that DNA can be used to perform computation.

Information theory, however, has so far done infinitely more for the understanding of DNA than DNA has contributed to the information sciences. For when meaning-deprived information theory, with its lifeless mathematical underpinnings, is coupled with old-fashioned biological expertise, which comes by way of rigorous experimentation, enormously powerful research tools are enabled. This coupling is the essence of an entire new field of research known as *bioinformatics*. It is also fundamental to the revolutionary changes that are occurring in the way in which biomedical research is conducted.

We've seen how genes have two personalities, one as a chemical composition and the other as an information carrier. RNA, on the other hand, is most closely identified with information alone. The RNA that encodes protein is known as *messenger* RNA (mRNA), for it carries a message from DNA to the protein-building apparatus. Next, consider amino acid sequences, the constituents of proteins, as information carriers. The amino acid sequence effectively tells the protein what shape to take and what chemical forces it should exert. No laws of physics are being broken here. Molecules are simply obeying physical forces, as are the electrons that race through phone lines. However, for each of these actions there is an underlying intent. Outside of living systems, when sunlight warms a rock or rain collects in a rivulet, for instance, there is no such readily discernible intent.

Like their nucleic acids antecedents, proteins appear two-faced. They are intricately folded chains of amino acids and they provide vital communication links among molecules. Growth hormone protein, for example, initiates a cascade of signals. The net result is cell growth and division, which appears on the macroscopic level as a gain in height and weight. Similarly, the protein leptin tells the body to slim down, and erythropoeitin tells the body to produce more red blood cells. In accordance to their role as information molecules, proteins have frequently been described as talking, tasting, smelling, touching, or seeing other molecules. Indeed, proteins do help sensations reach the brain by acting as signal relays. Cascades of proteins are put into action when photons impinge on the retina, fragrance lands on the nasal membranes, or sound shakes the delicate fibers in the inner ear. In all their roles, proteins and all other molecules of life can be considered as conveyers of information. Even so-called structural proteins convey messages. The proteins in your hair tell your hair to be kinky, wavy, or straight. It is by design.[1]

We have seen that the instructions for molecular structures have been transduced and digitized, and with the large-scale sequencing projects reaching genome closure it may appear as if the solution to many riddles of life are now before us. The central dogma of molecular biology is generally understood to mean that information flows from DNA to RNA to protein to trait.[2] Primitive bioinformatic tools readily translate DNA into RNA and RNA into protein (amino acid sequence) by simply applying base-pairing rules and the genetic code, respectively. Hence, it appears as if the first two legs of this flowchart can be faithfully replicated on a personal computer, and the only part of the puzzle that remains is the question of how proteins bring their messages to the macroscopic world. This impression, however, is very wrong.

Only select stretches of DNA are made into RNA, and only select portions of RNA are made into protein. Base-pairing rules say nothing about when and where RNA is made, and although the genetic code includes start (and stop) codons it doesn't predict which of the many AUG triplets will initiate the protein-building process or how often it will occur. Nonetheless, the instructions for regulating the production of RNA and proteins are themselves heritable and of great importance. (Recall that vertebrates share a largely similar set of genes and that spatial and temporal differences in gene expression are thought to differentiate the species.) Accessing this information cache has not been easy, but the lessons learned so far provide us with a basis for reevaluating the most common interpretation of the central dogma and for better examining the future prospects of bioinformatics.

Generally, codes can be deciphered by observing the relationship between the input symbols and the output symbols, deducing rules that explain these relationships, and if possible, testing the rules by manipulating the input and observing the effect on the output. Decades of this type of research have resulted in a growing body of knowledge of the rules that govern the beginning and end of RNA production (*transcription*), the splicing of RNA, and the start of protein production (*translation*). Several important insights have been made.

Firstly, these rules are often context-dependent. A DNA sequence that calls for the start of transcription in one organism may not do so in another. Furthermore, a sequence that directs the start of transcription in liver cells may not do so in skin cells. In other words, additional information is needed to determine whether a particular sequence will be made into RNA. The source of these inputs will be further described later.

Second, in any one cell many alternative sequences may direct transcription, RNA splicing, or translation. The RNA sequences AGGAGGU, CAGGU, GGAGG, UAAGGA, AGGAGGU, and a number of other variations each interact with the enzyme complex (a particular grouping of proteins, RNA, and other molecules) that initiates translation in bacteria. There is redundancy in the regulatory codes just as there is redundancy in the genetic code. Third, important information is contained in spacing. For example, the spacing between the two sequences AGGAGGU and AUG must be between three and seven nucleotides for it to signal the start of translation in bacteria. Spaces can be unexpected and thus contain information. Consider how a properly placed pause can make the difference between an uproariously funny joke and one that falls flat, or try removing the spaces between words in a sentence: Thespaceisaboutasimportantinwrittenlanguageasanysingleletter. (The spce is bout s importnt in written lnguge s ny single letter.)

Music also provides a good example of the information content of spaces. A silent moment in a musical composition may have tremendous effect. The composer John Cage is famous for his use of silence. One of his compositions (*4′ 33″*) features a lone pianist sitting silently for four and a half minutes (1952). There are, in fact, sounds in Cage's piece. The composition sounds different every time, for the audience hears different environmental sounds, such as those emanating from the audience itself. This introduces the final point about the rules that determine how information flows from DNA sequences to proteins, which is that idiosyncratic and random events play a role. Recall that mismatched base pairs are introduced during DNA replication. There are discrete sequence signals that guide the start of transcription and translation and determine where RNA is spliced, but like base-pairing, these regulatory signals are not absolute. The molecular machinery that reads these signals may be of limited supply in a cell and sequences will, in effect, compete against each other for access to them. Consider this issue in terms of probabilities. A stretch of 1000 nucleotides may be precisely replicated in 99.9 percent of occurrences. The chances that this sequence will yield a particular variation could also be calculated. Similarly, a stretch of sequence may have a 97 percent chance of causing transcription to initiate in a particular cell during a particular time period, while another stretch may have only a 73 percent chance, and a number of others could have just a 0.1 percent chance. These sequences could even be in front of the same gene. Some regulatory sequences speak louder than others do and some seem mostly inconsequential, like background noise.

Probabilities are also associated with the external inputs that influence transcription, splicing, and translation (the first point). Again, as with the gene-disease associations, we are confronted with bloody odds! But this doesn't mean we should despair in ever understanding how gene expression works. Just as a stronger physics was developed when the quantum mechanics and relativity of Erwin Schrödinger and Albert Einstein replaced Newtonian mechanics, replacing the solid arrows of molecular biology's central dogma with dotted lines may ultimately lead to more predictive and more useful life sciences. Furthermore, despite the inherent uncertainties, the system usually manages to work, in that it very often yields fully functioning organisms. If our bodies can thrive among such nonabsolutes, then so ought our conscious minds.

It is rather ironic that the predictive power of DNA sequences is often fogged in probabilities, for none other than Schrödinger suggested that "the entire pattern of the individual's future development" was encoded within the chromosomes, and that "in calling the structure of the chromosome fibres code-scripts we mean that the all penetrating mind . . . to which every casual connection lay immediately open, could tell from their structure whether the egg would develop, under suitable conditions, into a black cock or a speckled hen."[3] Schrödinger's statements, given in 1943 in a series of lectures entitled "What Is Life?" provided inspiration and conceptual guidance to Crick and cohorts in their efforts to discover the structure of DNA and characterize its actions. However, despite his great influence, Schrödinger's musings on biology were admittedly naive. His unmatched brilliance was displayed decades earlier when he elegantly proved that many properties of sub-atomic particles could only be described in probabilistic terms!

Information theory grew out of the mathematics of probabil-ity, and new bioinformatic tools exploit statistics mercilessly. Thousands of transcription and translation start sites and splice sites have been determined through laboratory research, and now bioinformaticians can utilize Hidden Markov Models, neural nets, or a host of other statistical innovations to construct algorithms that ferret out this information in fresh genomic sequences. For example, when the gene-finding program GeneMark, developed and improved upon for over a decade by Mark Borodovsky, James McIninch, and colleagues at the Institute of Molecular Genetics in Moscow, was applied to 10 fully sequenced bacterial genomes, the program correctly identified known bacterial genes 97.3 percent of the time, and for 78.1 percent of these it correctly identified their starting codon.[4] Currently, GeneMark and other gene-finding programs are much less successful with human genomic DNA. They can identify about 50 percent of the genes, but as more experimental data is collected such programs are likely to improve. Other computer programs specialize in identifying splice sites, finding translation starts, and other sequences that have been deemed meaningful by researchers. One can even search blindly for frequently reoccurring sequences. Such sequences must have "meaningful" features that have enabled their survival.

Beyond sequence and structure lies function, the Holy Grail for gene hunters. In the early days of molecular genetics genes were identified on the basis of their function, but when sequences began gushing into the databases like water from a hydrant, desig-nations of function began to lag. Currently, there are many thou-sands of these orphaned genes, poor unchristened protein-coders that are nonetheless rich in concealed information. Fortunately, many would-be parents are eager to adopt. The assignation of even a putative gene function provides the assignee with a head start on further research and a much stronger basis for patent protection.

Of course proteins often have multiple functions, and function can be elucidated at multiple levels. Not surprisingly, a new research focus and a new industry have emerged just to tackle questions of gene function. In "functional genomics," innovative technologies are aimed at rapidly and more effectively assigning gene functions to sequences. Bioinformatics provides many of the speediest and most cost-effective means of assigning a putative gene function. Once again, both a wealth of raw sequence data and the knowledge gained through years of experimentation can be skillfully leveraged through the application of information theory.

A high level of sequence relatedness implies a common ancestor and at least some degree of functional relatedness. The functional similarity may originate from a particular domain or motif consisting of a set of related sequences that constitute a portion of a protein. How restrictive these sets are will vary from one functional domain to another. Particular amino acids that provide chemical reactive sites or structures critical to the protein's specific actions may exist in various regions of the linear sequence of amino acids. Rules that define their spacing and composition are complex and probabilistic. Nevertheless, information sciences allow these rules to be determined, and once the rules are known, search engines such as BLAST can quickly detect such sequence similarities and thus make tentative assignations of protein function.

For example, there is a large family of proteins known as *zinc-finger proteins*, which bind to DNA and act in the regulation of gene expression. Each zinc-finger protein contains a repeated series of amino acids that fold into a structure that is stabilized by a zinc ion. Years of experimentation helped in deducing that each zinc-finger domain consists of Phe/Tyr-X-Cys-X_{2-4}-Cys-X_3-Phe-X_5-Leu-X_2-His-X_{3-5}-His (where each three-letter code represents a specific amino acid and X_n signifies a series of any n amino acids).[5] Through the application of bioinformatic programs, hundreds of new members of this important family of proteins have been discovered within large-scale sequencing databases.[6]

A protein encoding sequence is a stepping stone for additional information analysis. Computer programs may predict the physical structure that a particular sequence of amino acids assumes and the chemical properties that it will have. The sequence tells the protein what shape to assume, and, for a class of molecules whose interactions have been likened to that of a key fitting into a lock, shape can be very important. AIDS researchers listened carefully to the sequence of HIV protease. Knowledge of its three-dimensional structure helped in the design of lifesaving new drugs, molecules that interfere with the HIV protein not just by fitting into the keyhole but by jamming it as well. The structures of hundreds of proteins have been determined by analyzing the pattern of X-rays that diffract through crystallized forms of the proteins. Knowledge gained by these studies is used to formulate structure-predicting programs. As with the human gene-finding programs, structure-predicting algorithms are less than perfect. They, too, are probabilistic, but their predictive powers are improving because of additional sequence information, better experimentally derived knowledge of rules, and ongoing mathematical innovations.

One finds information virtually everywhere among the molecules of living organisms. Not only does DNA talk to RNA and RNA talk to proteins, but proteins speak to each other and to both DNA and RNA. Zinc-finger proteins tell specific regulatory sequences to turn on transcription, for example. Besides the nucleic acids and proteins there are many additional chemical messengers, such as steroid hormones and neurotransmitters that relay messages to receptor proteins. Often the message relay system is not linear.

For example, leptin receptors in certain brain cells activate sets of proteins that relay a "decrease appetite" signal, while other protein messengers are sent to increase energy expenditure. Reproductive tissues may alter their rate of development in response to leptin, while leptin-producing cells may halt production of additional leptin, and additional yet-to-be-discovered recipients may perform other acts. (The intended use of leptin as a fat-reducing drug is complicated by these diverse messages. Many promising new drugs have been cut down by unintended side effects, the consequence of undesired messages.)

The bifurcation of signals, feedback, and cross-talk is common. Thus, rather than a row of dominos, the message relay system resembles a very intricate network or web. It is for this reason that the linearity of the central dogma fails to provide an adequate framework for understanding the full scope of information flow, and the genome should not be thought of as the pinnacle of an information hierarchy. Instead, the sequences of the genome are in a multiway conference call with the molecules that surround them. Not only were sequences and other molecules in such a state when we were conceived, they have probably been online since the start of life. Welcome to the Internet? Heck, it's been in us since the dawn of our creation!

Although molecular messages are passing every which way both within and between cells, there is, of course, an order to their commerce. It has often been said that to understand the language of life one must do more than learn the sequence of letters or identify all the words. In other words, knowing the sequence of the entire genome and all the genes will be insufficient. One must understand how small messages join together to create larger and more meaningful messages. To understand the molecular basis of life and of disease will require knowing how genes, regulatory sequences, RNA molecules, proteins, and other molecules work together to generate nerve tissue, fight a bacterial infection, maintain liver function, and so on. In linguistic terms this means understanding the grammar of life.

The way in which the molecules of life communicate has been likened to the way in which people communicate, so much so that linguistic terminology abounds in molecular biology. Nucleotides are known as *letters*, triplets that encode amino acids have been called *words*, collections of genes are known as *libraries*, proteins are *translated* from nucleotide sequences, and proteins *talk* to each other. The goal of the molecular biologist in the genomic age has been described as translating the *language* of the cell. The Human Genome Project has been likened to the Rosetta Stone, the piece of rock whose inscriptions in three ancient scripts enabled the Egyptian hieroglyphic code to be broken. The human genome itself has been called the *Book* of Man. The genome with all repetitions, space holders, and nested meanings has also been likened to a poem (an analogy undoubtedly favored by defenders of so-called junk DNA). A music analogy also has some adherents. Several experimentalists have even represented the four different nucleotides as notes and used DNA sequences as musical scores. But the resemblance to language has been more compelling and has been strong enough for several researchers to probe biological questions within the framework of linguistic theory. For example, Julio Collado-Vides of the National Autonomous University of Mexico has derived a minimal set of grammatical rules that describe the regulation of a particular class of bacterial genes. The regulatory sequences, genes, and gene product are like verbs, adjectives, and nouns, and their combinations, the set of possible sentences, define the grammatical rules.[7]

The network of molecular messages within an organism also resembles electronic circuitry. Regulatory sequences act like switches that turn on or off the production of RNA. These switches process information; they receive input messages from their surroundings, mostly from proteins, and respond with an output message, gene expression. The French biologists Jacques Monod and Francois Jacob were the first to characterize a genetic switch. Monod and Jacob wondered how E. coli "knows" to turn on the gene for beta-galactosidase enzyme when and only when lactose sugar is present in its environment (which happens to be human intestines). The beta-galactosidase protein digests lactose sugar, thereby liberating energy and carbon that the bacteria need to live. In the mid-1960s Monod and Jacob found that a type of protein, which they termed a *repressor,* binds to DNA upstream of the beta-galactosidase gene, preventing transcription from occurring. Lactose molecules tell the repressor (by binding to it and thereby changing its shape) to fall off the DNA and allow expression of beta-galactosidase to occur. This type of switch or some variation of it exists for all genes. Some involve activator molecules (usually proteins like the zinc-finger proteins); others utilize repressors; and many use complex combinations of activators and repressors.

Genetic switches may be arranged in a circuit, where the products of one switch regulate another switch whose products regulate several others and so on. In 1963 Monod and Jacob suggested that cell differentiation, the enigmatic process by which one cell type gives rise to another, may be controlled by such circuits.[8] From this idea sprang the dream of understanding cell differentiation through molecular circuit models. The circuitry of even a small network of molecules can be quite complex, yet it can be delineated. This was elegantly demonstrated in the 1970s by Harvard professor Mark Ptashne when he elaborated the complex circuitry of a relatively simple bacterial virus.[9] Decades later a team of scientists used the language of electronics (capacitors, transistors, and the like) to encode the circuitry of this network of about 20 molecules and DNA sequences.[10]

Beginning in the late 1960s, Stuart Kauffman, then a medical student and now a New Mexico–based author, entrepreneur, and biology theorist, pioneered the development of mathematical models of even more complex molecular circuitry. In his models of living systems Kauffman employs a Boolean network, where thousands of genes are represented as elements that are either on or off, and the on/off status of certain elements determines the status of other elements by the application of an element-specific formula that applies the logic functions AND, OR, NOT, etc.[11] For example, the formula may state that gene A will be on if and only if gene B is on and gene C is off. Gene B, on the other hand, will be on only when both genes C and D are on, unless both genes A and E are both on.

Kauffman's models are dynamic (the on/off states of the genes go through a series of changes over time) and may mimic the transformation of primordial (or stem) cells into terminally differentiated cells (such as mature liver, bone, or muscle cells). If each gene were represented by a light bulb on a grid, then the grid might cycle through a pattern of changes before settling into a repeating pattern, which represents the genes in a terminally differentiated cell. An environmental insult or a genetic defect can be modeled by altering either an element's on/off status or its formula for switching states. Remarkably, Kauffman's models mimic cells' curious blend of stability and vulnerability upon perturbation. Certain perturbations would send the light bulb grid into a brief change in light pattern, only to return to its repeating pattern of lights. Other perturbations would send the light bulb grid into unending chaos (like a cancer cell) or into a fixed pattern (akin to cell death).

Kauffman did not describe models of actual gene networks. His Boolean networks were entirely theoretical. Furthermore, real genes are not simply on or off. They are expressed at a functionally relevant level within a gene-specific range of levels. Nonetheless, the properties of his Boolean networks were so intriguing and life-like that they helped inspire a field of study known as *complexity theory*. They also led Kauffman to speculate that certain networks may self-organize, spontaneously giving rise to life.[12]

Such theories are fascinating, but what about modeling real molecular networks, solving the riddles of complex organisms, and finding new treatments for diseases? This may not only require sophisticated mathematics and advanced computer systems; it may also require accurate and extensive access to actual molecular information. This was the sentiment that in the early 1980s spurred Norman and Leigh Anderson to campaign for support for a Human Protein Index project. At that time Leigh Anderson and coworkers wrote that information-based solutions to the riddles of cellular differentiation are "likely to become feasible within the next decade, if a large enough base of information is assembled."[13] Sure enough, by the late 1990s real gene data sets were becoming extensive enough for network modelers to use.

Each January many of the world's brightest bioinformaticians gather in Hawaii at the Pacific Symposium for Biocomputing. It is a great opportunity for biologically inclined mathematicians and mathematically inclined biologists to both exchange ideas and test their luck at surfing. In 1998, when I attended this event, I felt that a significant transition might be underway. I heard several reports on theoretical molecular network models, including a technique to deduce the on/off formulas of Boolean networks (a process known as *reverse engineering*) and designs for non-Boolean networks that allow for a continuum of gene expression values.[14] More important, I heard reports on models that attempted to use real biological data captured using newly developed genomic technologies. It also seemed noteworthy that this traditionally academically orientated conference was attracting increasing numbers of industry-employed scientists.

In summary, information processing occurs throughout the molecular network. Molecules seldom simply relay messages. More often, they receive input messages from several different sources, apply their particular formula, and then send one or more output messages. This creates the branches and bifurcations in the network. Much, though certainly not all, information processing occurs at gene regulatory sequences. Active genes are continually replacing the proteins of the cell and modulating the protein levels in response to various signals. Ten thousand or so genes may be active simultaneously in any human cell, with the set of expressed genes varying from cell type to cell type (as was described in Chapter 6). By the language analogy, the set of expressed genes reflects the genome's contribution to an ongoing conversation between the genome and its surroundings. The pattern of gene expression can thus reflect the state of the cell, its health and well-being, and its response to disease and disease treatments. Information is also processed by other molecules within the cell. Proteins, in particular, often act as dynamic switches, gatekeepers, or amplifiers, as well as relays. The access and interpretation of this information is critical to the development of a molecular understanding of life and its anticipated benefits.

The importance of gene expression was determined through decades of laboratory research. Assays detect RNA levels through hybridization (base-paring) to gene-specific probes (single-stranded DNA or RNA complementary to the RNA being assayed). Protein levels are detected using antibody molecules, which are proteins that bind only to very specific fragments of proteins or other molecules. The application of these techniques has been so widespread that they are documented in well over 100,000 scientific publications. As with DNA sequencing, however, these assays have traditionally been done a single gene at a time. Just as EST technology opened the floodgates to the information content of the genome (i.e., the sequences), other technologies have opened up access to the information output of the genome (i.e., the gene expression levels).

These new technologies and the way they came about are integral to the current revolution in the biomedical sciences. They lay the groundwork for understanding life not in terms of individual genes and gene functions, but rather in terms of complex molecular networks. They are opening the door for a new approach to treating disease that intervenes in these complex molecular networks when they go awry, instead of trying to treat individual genes.

14

Accessing the Information Output of the Genome

In the early 1980s Leonard Augenlicht, a cancer researcher with appointments at Albert Einstein University and Montefiore Medical Center (both in New York City), was looking for insight into the genomewide changes that occur in cells as they transform into cancerous cells (a process that can be mimicked with cells grown in an incubator). Augenlicht knew that a complex pattern of changes occurs during the transformation process. Years earlier, researchers had noted that between disparate cell types 15 to 25 percent of the mRNA transcripts differed and among the transcripts that were shared there were widespread differences in the level of expression. Cancerous cells and their noncancerous precursors were thought to also differ in their overall pattern of gene expression. In a 1982 article in the journal *Cancer Research*, Augenlicht and coauthor Diane Kobrin noted:

> Analysis of the expression of randomly selected clones in a variety of tissues and tumors is of value for 2 reasons: (a) It allows the identification of cloned sequences which can be used as probes in studying the nature and mechanism of change in gene expression in carcinogenesis; (b) a compilation of the data can reveal interesting patterns and shifts in expression for individual sequences and subpopulations of sequences.[1]

Thus, well before either Craig Venter or Randy Scott had begun large-scale sequencing of randomly chosen genes, Augenlicht and Kobrin had concocted a way of simultaneously monitoring the expression of many such genes.

149

Augenlicht and Kobrin took 400 colonies of bacteria, each carrying a randomly selected gene from a tumor cell cDNA library, extracted the DNA, and spotted it in an ordered array on multiple copies of a paperlike membrane. These membranes were then used to measure the expression level of these genes in any number of cell or tissue samples. For each of several different mouse tumor and nontumor cell lines, mRNA was extracted, labeled with a radioactive chemical group, and allowed to hybridize to the spotted DNA on the membrane. The membrane was washed, exposed to radiation-sensitive film, and then scanned using an imaging device that recorded the intensity of each spot. In this way, gene expression levels in mouse tumor and normal cells were compared, not just one gene at a time, but hundreds of genes at a time.

Augenlicht did not stop there. He and his cohorts then made arrays of 4000 human genes and used them to interrogate human cancer cells. Data points from these experiments went directly from the imager into a computer, allowing the digitized gene expression information to be readily analyzed. Five percent of the arrayed genes showed a threefold or greater increase in expression in colon cancer cells relative to normal colon cells. Three percent of the genes decreased to a third or less. There appeared to be distinct patterns of gene expression changes at each stage of colon cancer development. Furthermore, the colon cells from patients with a particular inherited form of colon cancer had gene expression patterns that differed from others. People differ in their risk profiles for cancer in accordance with their lifestyle and genetic makeup. Colon cancers differ in their progression and response to treatments. It is clear now that Augenlicht was onto a powerful new technology to detect these differences at the molecular level. The technique had potential not just in research, but also in the diagnosis and treatment of cancer and other diseases.

However, in 1986 the science authorities who reviewed Augenlicht's work for the journals *Science* and *Nature* were not sufficiently impressed. After all, the spots on the membranes represented unidentified genes. Even if they were sequenced, which Augenlicht did in a few cases, the function of many of these genes was entirely unknown. They were among hundreds or thousands of genes whose expression level changed. Furthermore, the significance of a 3-fold or 1.2-fold or 10-fold change in mRNA level was not at all clear. His report was rejected by the high-profile journals, though it was published in *Cancer Research*, a publication that is highly regarded among cancer specialists. Nonetheless, Augenlicht's report received scant attention. Even worse, an application for funds earmarked for further development of the system was rejected by the NIH. Augenlicht had hoped to address some of the lingering biological questions as well as to improve the computer system, which at the time could only handle one-tenth of the data points.

The response was very similar to what had greeted the Andersons and coworkers at Argonne for their protein index project and to the one that Craig Venter would later receive for his EST project. The idea of letting randomly selected molecules direct the course of research was simply unpalatable to most scientists. However, Augenlicht's work did manage to convince a patent examiner of the novelty and usefulness of his invention. A U.S. patent was issued in 1991, but it, too, would gather dust. Years would pass before the sky's-the-limit entrepreneurial spirit and "faster, better, cheaper" engineering recast Augenlicht's apparatus in the form of the "gene chip" or "microarray."

Far from Augenlicht's Bronx laboratory was a utopia for good ideas, a place where know-how, imagination, and money run so deep that it seems as if nothing can stop a good idea (and even many not-so good ideas) from being thrust into commercial development. Silicon Valley, the area surrounding the southern part of the San Francisco Bay, is where sophisticated electronic instrumentation, silicon transistors, the personal computer, Hewlett-Packard Inc., Fairchild Semiconductor Inc., and Apple Computer Inc. began. As a result, the world will never again be the same. And this area is also where the technology to monitor the expression of thousands of genes simultaneously was first commercialized.[2] It is worth considering the social influences exerted here, because events in Silicon Valley are beginning to have a profound impact on the biomedical sciences.

Like few other places in the world, the San Francisco Bay area has never really been dominated by a single ethnic population.

Native Americans, Spaniards, Russians, Mexicans, Chinese, various Europeans, and all sorts of contemporary Americans have all called the area home, but no one culture has ever been effectively enforced. From this stew of ethnicities emerged a culture of tolerance and mutual respect. The high-tech industry of Silicon Valley is overlaid on this Northern California culture, represented by a multiplicity of temples, synagogues, communes, churches, covens, bathhouses, monasteries, outdoor wonderlands, and other focal points. Beneath the compulsive drive for "faster, cheaper, better" is a pervasive hippie culture, which took root in Northern California well before the term was invented in the 1960s and still exists to some extent today. Fruit trees and organic gardens are still common, and the smell of marijuana still lingers in the air. Casual, yet cohesive, the social culture acts as lubricant for the flow of ideas, a foundation for the building of trust, and a catalyst for deal-making.

Stanford University owns more than 8000 acres of Silicon Valley. University founder Leland Stanford once maintained a horse farm on some of this expanse. Now "The Farm," as it is still called, raises businesses. In its stables are many of the high-tech thoroughbreds that are credited with transforming the global economy and thrusting the world into the Information Age. Not only is a portion of Stanford's land leased to high-tech companies, but ideas and people from throughout Stanford seed the industry. For example, an effort by staff members to link various computer systems at Stanford was transformed into the enormously successful Cisco Systems Inc., and two Stanford graduate students' personal interest in cataloging sites on the then-nascent Internet was transformed into Yahoo! Inc.

Facilitating these transformations is the world's highest concentration of venture capital firms, most located in Menlo Park, less than a mile away from Stanford. These venture capital groups, the capital-rich venture departments of area high-tech firms, and a slew of wealthy and technology-savvy individuals manage to commercialize ideas out of the many area research labs, no matter whether they are housed at a university, an established company, a start-up, or in a garage. When Mark Andreeson arrived from the Midwest to work for a local firm, venture capitalists swooped down and within a few months transformed his ideas for viewing

Internet sites into Netscape Inc. Numerous such outfits have also made a significant impact on society. Among the most notable are Oracle Systems, which brought innumerable organizations into the Information Age with its relational databases, and Intel Inc., the premier maker of central processing units, the computer chips that are most integral to the machine.

Chips are the subject of Moore's Law and the embodiment of an industry based on dense circuitry etched in metal and silicon. When Augenlicht's idea was recast as "gene chips," the full entrepreneurial force of Silicon Valley was activated (see Fig. 14.1). Only a few discoveries and barely a cent had been earned from gene expression chips when President Bill Clinton made reference to them in his State of the Union Address in January of 1998, but his comments serve as a testament to the upwelling of interest:

> And while we honor the past, let us imagine the future. Think about this—the entire store of human knowledge now doubles every five years. In the 1980s, scientists identified the gene causing cystic fibrosis—it took nine years. Last year, scientists located the gene that causes Parkinson's disease—in only nine days. Within a decade, "gene chips" will offer a road map for prevention of illnesses throughout a lifetime. Soon we'll be able to carry all the phone calls on Mother's Day on a single strand of fiber the width of a human hair. A child born in 1998 may well live to see the 22nd century.[3]

Clinton was acting as a soundboard for the millions of U.S. citizens who elected him. There is no doubt that there is a strong desire to direct the powerful fast-paced engines of the computer and communication industries towards an enemy common to all humankind. It is high time that the full brunt of innovative powers, including those of science, engineering, law, and economics, be applied to the scourges of humankind, to do as Clinton suggested, to "begin a revolution in our fight against all deadly diseases."[4] President Nixon had roused many in 1971 with a declaration of war with cancer. Twenty-seven years and a millennium forward there is a strong sentiment that a new America, one more at ease socially

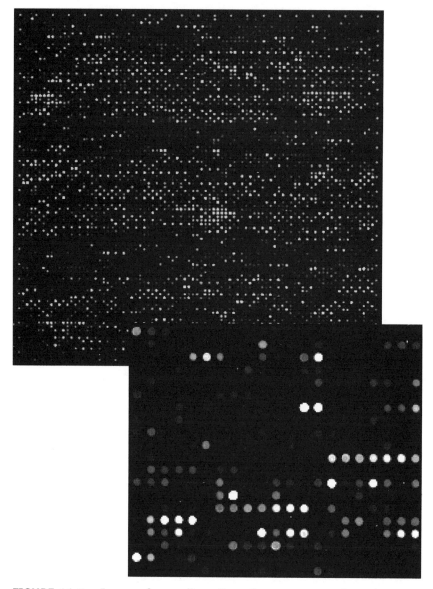

FIGURE 14.1 Image of a small portion of a gene expression microarray (DNA chip). Distinct gene sequences were synthesized at each spot by successively depositing individual nucleotides via ink jet printing. The microarray was then hybridized to labeled samples of mRNA, washed, and read by a scanner. *(Microarray image courtesy of Agilent Technologies.)*

and triumphant over communism, high unemployment, interest rates, and inflation, ought to be better able to wage war with the molecular demons that are thought to underlie our ills.

The inspiration for Clinton's comments on "gene chips" was most likely GeneChip™, which was at the time a new commercial product from Affymetrix Inc. of Santa Clara, California. Affymetrix grew out of Affymax Inc., a Palo Alto company that had been try- ing to automate the chemical synthesis of drug candidates. Affy- max scientists and engineers were seeking to manipulate and keep track of tens of thousands of molecules at a time.

At a meeting of Affymax's scientific advisors in the late 1980s, Leighten Read, a consultant, suggested the application of a process used in the semiconductor chip industry.[5] On the path towards "faster, cheaper, better" computers, semiconductor chip manufac- turers adopted the use of light as a means of bringing energy to specific places on the silicon wafer, thereby directing the chemical etching of desired circuitry. The process, called *photolithography*, utilizes a stencil-like mask that allows light to only reach very spe- cific places on a reactive surface.

With the help of Michael Pirrung and Lubert Stryer of Stan- ford, the idea of applying photolithography to chemical synthesis took root at Affymax.[6] Steve Fodor, then a postdoctoral fellow at the University of California at Berkeley, was recruited to lead the effort, which was initially directed towards the synthesis of pep- tides. In February of 1991 in a report featured in the journal *Sci- ence*, the Affymax team described the light-directed synthesis of a chain of five amino acids at discrete regions on a glass slide. In the same report the researchers also described the formation of chains of two nucleotides, and it would be this application, the creation of arrays of synthetically made DNA, that would yield the team its first commercial product.[7]

Affymetrix was spun off of Affymax in 1993, with Fodor as the chief executive officer. In December of 1996 the use of a gene expression chip was demonstrated in a report published in the journal *Nature Biotechnology*.[8] Gene sequence information was used to design sets of 20-nucleotide-long probes that were each complementary to one of 118 mouse genes. These probes were synthesized on glass using the photolithographic method, and the resulting oligonucleotide arrays were hybridized to fluorescently labeled samples of cDNA derived from a variety of different mouse tissue sources. The intensities of the signals were used to compute the expression level of each of the 118 genes in each of the tissues. The ability to detect low levels of gene expression and the potential to interrogate a far greater number of genes created a buzz among both scientists and investors. Affymetrix also made chips that could very accurately detect most sequence variations in several important disease genes, including p53 and HER2, which play important roles in many cancers.

One of the forces that propelled the development of gene chips or *microarrays*, as they are often called, was the desire for more sequence data. In 1988 the DNA sequencing world was abuzz with a new sequencing scheme called *sequencing-by-hybridization* (SBH), for in the fall of that year Radomir Crkvenjakov and Radoje Drmanac of the Institute of Molecular Genetics and Genetic Engineering in Belgrade, Yugoslavia, presented this innovative concept at an international sequencing meeting in Valencia, Spain. At least five groups left a historical impression by publishing related schemes either in scientific journals or in patent applications, including groups led by Andrei Mirzabekov of the USSR, William Bains of the United Kingdom, Stephen Macevicz of the United States, and Edwin Southern of the United Kingdom (Southern had previously developed the forerunner of all hybridization-based assays, a DNA detection technique known as the *Southern blot*).

There were many variations on the SBH scheme, but the fundamental idea is as follows: An ordered array of every possible nucleotide sequence of a particular length (typically between 5 and 10 nucleotides) is created. An unknown sequence is radioactively or fluorescently labeled and made to hybridize to each complementary sequence on the array. The set of sequences that hybridize to the unknown sequence is identified and represents every fragment of the unknown sequence of a particular length. A computer program then takes these sequence fragments, finds the order in which they overlap, and thereby electronically reconstitutes the unknown sequence (see Fig. 14.2).

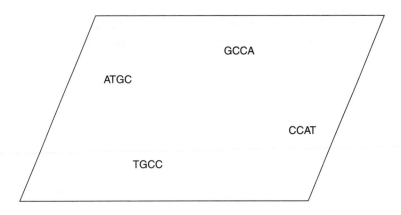

```
        A T G C
        T G C C
        G C C A
        C C A T
    _____
        A T G C C A T
```

FIGURE 14.2   An example of sequencing by hybridization (SBH). (a) All possible oligonucleotides (short stretches of synthetically made DNA) are spotted on a two-dimensional array. An unknown sequence is hybridized to the array, resulting in four "hits." (b) A computer program deduces the sequence by aligning the four sequences that hybridized to the unknown sequence.

The SBH concept proved to be quite difficult to put into practice, but of all the efforts the Drmanac and Crkvenjakov team made the most headway. The two researchers had come to Silicon Valley, where they cofounded HySeq Inc. Curiously, the defunct government of war-torn Yugoslavia initially owned a significant stake of the company, for it held the original rights to Drmanac and Crkvenjakov's invention. HySeq chips have been used to identify tens of thousands of genes and gene variants and they therefore represent yet another successful transition into large-scale multigene research. Along the way, the SBH idea spurred advances in arraying technology, imaging, and software, all of which was necessary to deal with tens of thousands of data points. This in turn contributed to the use of DNA chips for large-scale gene expression analysis.

Incyte began tapping into gene expression information at the outset of its high-throughput sequencing program. A statistical analysis determined that by sequencing 5000 randomly chosen ESTs, a valid measure of gene expression could be achieved for genes expressed at middle or high levels (arbitrarily >3 in 5000) in any particular group of cells.[9] A tally of ESTs from the same gene (clustered ESTs) can be readily converted into a percentage abundance and stored in a database for further analysis.[10]

A list of expressed genes and their percent abundance is known as a *transcript image*. It is a snapshot of the genome's activity in a particular tissue at the time in which the RNA was isolated. Incyte's relational database stored hundreds of transcript images and allowed the informational output of the genome of different cell or tissue types to be compared in innumerable ways. For example, genes overexpressed or underexpressed in prostate tumors relative to nondiseased prostates could be readily identified with a few clicks of a computer mouse. Among these was the gene for Prostate Specific Antigen (PSA), a protein that is released from prostate tumor cells into the bloodstream. Testing for PSA enables early diagnosis of prostate cancer, which allows early treatment and the preservation of many lives. A more sophisticated, statistical analysis of Incyte's database uncovered a dozen genes, including several novel genes, which had a pattern of expression similar to that of PSA.[11] It is hoped that these genes may enable still better diagnostic tests and new avenues of research towards a cure.

Gene expression comparisons using EST databases work well for the scores of middle- and high-abundance genes, but in order to accurately track changes in expression of the hundreds of low-abundance genes one must either pick far more than 5000 sequences or combine transcript images from many similar tissues. The 5000 randomly chosen cDNAs are derived from a pool of a million or so mRNAs (the components of a typical cDNA library), so comparing an abundance of zero or one in 5000 to one or two in 5000 is statistically not very meaningful.

However, the messages these low-abundance genes bring to the body may be very meaningful. Receptor proteins, for example, may be of extreme low abundance, yet subtle changes in their expression may lead to big changes for the cells in which they reside. They often act as information gatekeepers at the surface of cells; they are very sensitive to input signals and able to initiate a large-scale coordinated response. Changes in receptor protein levels are usually mirrored by changes in receptor mRNA levels. In order to better access this type of information, Incyte scientists and engineers evaluated dozens of different options and then settled on two strategies. They continued to sequence more libraries, sequencing some to a depth far greater than 5000, and they gained access to a highly sensitive low-cost microarray technology. This technology came from yet another Silicon Valley success story.

Dari Shalon and Professor Pat Brown of Stanford University came up with the idea of spotting gene fragments at high density onto glass by way of robotically controlled sets of specialized metal pins. The metal pins are like miniature fountain pens, capable of pulling up tiny droplets of DNA from plastic wells and then delivering them to precise locations on a glass "chip." These microarrays of gene fragments were hybridized to two sets of cDNAs, each from a distinct tissue or cell source and each labeled with a different fluorescent dye. At each spot the relative signal of one dye compared to the other reflects the relative level of expression of that particular gene.

Upon graduating from Stanford in 1994, Shalon started Synteni Inc. in Fremont, California, with the help of local venture capital firms. Synteni scientists and engineers advanced the technology so that the chips could be sold commercially. Incyte purchased Synteni in January of 1998 and immediately began arraying gene fragments from vast collections of DNA amassed through years of high-throughput sequencing.

Dozens of other groups, including new ventures, academic laboratories, large pharmaceutical companies, scientific equipment makers, and prominent semiconductor and computer makers, also started gene chip programs. In addition, several other types of genome-scale gene expression assays were developed.

In this society we step into the future by tugging on our own bootstraps. The medical benefits to be derived from this trove of information are largely unknown, yet the prospects for such large-scale access to gene expression information have captivated a great number of people. Hundreds of millions of dollars and millions of hours have been invested in efforts to access this data. The efforts have been hyped, touted, and sensationalized. Fortunes have been made and larger sums of money have been put at risk. Information technology and engineering powerhouses, notably Motorola and Agilent Technologies (a huge Hewlett-Packard spin-off), have entered the field. Feathers have been ruffled, egos have been bruised, sabers have been rattled, and lawsuits have been filed.

Meanwhile, back in the Bronx, Augenlicht simply continued to pursue his passion. For a long while he put the array project on the back burner and applied more established avenues of research to a handful of colon-cancer-associated genes, tracing the chain of molecular events back to their root causes. In particular, he linked a number of important molecular changes in colon tissue to diets high in certain fats. Had Augenlicht, by some twist of fate, fallen into the Silicon Valley vortex in the 1980s the history of gene chip development could have been very different, but it is hard to even guess as to whether the world would have been better off.

15

The Genomics Industry

The region south of San Francisco is also heralded as Biotech Bay, Silicon Valley's younger and less prominent sibling. At Stanford University in 1973, Stanley Cohen and Annie Chang cloned the first bit of DNA, splicing a gene from one type of *E. coli* bacteria into a circular piece of DNA which was capable of replicating in other *E. coli*. The following year they helped introduce a toad (*Xenopus laevis*) gene into the same circular piece of DNA. Cohen and Herb Boyer of University of California at San Francisco (UCSF) received credit for conceiving the idea, and Stanford's technology licensing office became renowned by making the recombinant DNA technology broadly available in nonexclusive deals. This move would garner UCSF, Stanford, Boyer, and Cohen about $200 million over the 17-year life of the patent. One of the early licensees was Genentech Inc., the first company to make a commercial product from the new gene-based technology and the company widely recognized for initiating the biotechnology industry.

The biotechnology industry is said to have begun over beers at a 1976 meeting between Boyer and a young venture capitalist named Robert Swanson. In the courtyard of Genentech's research labs in South San Francisco a life-size bronze statue of the two sitting at a table with drinks in their hands commemorates the moment. The company was started that year, and Genentech researchers quickly figured out how to make bacteria produce medically useful proteins using recombinant DNA technology. They then scaled up the laboratory procedures into an industrial-size enterprise, one whose products were pure and safe enough to meet the rigid requirements of pharmaceutical regulators. Genetically engineered human insulin reached the market in 1982, only six years after Boyer and Swanson started Genentech.

In the years that followed, a dozen or so new biotech outfits were spun off of Genentech; dozens more were born from mama Stanford; and more significantly, several thousand such companies arose throughout the world. By 1997 over 1200 biotech companies existed in the United States alone and together they employed over 140,000 people.[1] Most of the new enterprises aimed to develop better medicines. (Others were involved in agriculture and industrial chemicals.) A cornucopia of new drugs was proposed, enough to fulfill virtually every imaginable unmet medical need. Herein lies a great chasm. On one hand, businesses naturally inflate their prospects; they artfully promote themselves so as to pique the interest of investors and clients. Biotechnologists and their backers have become skilled in this art. On the other hand, the rate at which biotech companies have delivered the goods has been rather abysmal. For example, on over 21 occasions clinicians administered to patients new experimental drugs designed to ward off sepsis, an often deadly condition brought on by bacterial toxins. In each case the drug failed to be safe enough and effective enough to warrant regulatory approval. The biotech industry began more than 20 years ago, yet genetically engineered drugs have supplanted only a tiny fraction of the entire drug arsenal, and for every genetically engineered insulin there have been hundreds of costly and time-consuming failures.

The allure of biotechnology has attracted many investors, but most have been bruised and battered by the experience. Throughout the 1980s and 1990s the U.S. stock market, and in particular the market in high-technology securities, was manic. The rising tide raised nearly all boats, with the biotechs being a notable exception. An index of the leading biotech stocks (symbol ^IXB) rose only 24 percent over the three-year period from July 1996 to July 1999, a period of time in which a comparable index of leading pharmaceutical company stocks (symbol ^DRG) rose 150 percent, while both the Dow-Jones industrial average and the technology laden NASDAQ index each climbed over 100 percent. Until the year 2000 the stocks of the vast majority of biotech companies had floundered. A few companies went out of business entirely. Most survived by what has been referred to as an "arsenal of financial gymnastics."[2] Ownership was shuffled, new investors were brought in, assets were transferred to other companies, liabilities were deferred, and more often than not, share prices only stagnated or declined. Quick-to-the-exit insiders and some astute short sellers have been able to deftly profit from the inflated hopes and hype that seem to plague the industry. Other investors have been left holding the bag. "Because biotech is tricky," the heading of one investment consultant's Web page appropriately warns.[3]

Only about 20 out of 1200 biotech companies in the United States have product revenues.[4] Combined product revenues for all U.S. publicly traded biotech shops amounted to only about $17 billion in 1998, compared to over $25 billion for Merck alone, an established drugmaker, or over $20 billion for Dell Computer, a company that began only nine years earlier. In terms of profits the biotech industry appears even worse. The industry as a whole has lost money in each of its 30 years, and profitable biotech companies can be counted with a single hand. Some of the better outfits, such as Genetics Institute, Chiron, Immunex, and Cetus, have either merged or have been bought out by mightier partners. Genentech came close to becoming the first home-grown fully integrated drug company; but it appears that management may have lost their taste for financial risk, for Genentech has come to rely more and more on the drug discovery research of others and the marketing and sales infrastructure of the established drugmaker Roche Inc. By aligning with Roche, Genentech shareholders received the security of a stay-at-home domestic partner, with an older, more powerful, and worldlier mate. And this is the crème de la crème; overall the biotech industry's legs have been weak. It has lurched forward, but only with regular injections from established pharmaceutical makers and booster shots skimmed from the overflow of capital-rich markets.

Given the high failure rate, the losses, and the low returns, one wonders what kind of fools would invest in biotechnology. And given the unpredictability of the science, the vagrancies of the business, and the low pay relative to other high technology jobs, who would be crazy enough to stake a career in biotechnology? No doubt, there are many motivations, but a few themes standout. For some people risk is part of the draw. Despite the poor odds, the prospect of finding gold, whether it is a cure for cancer or pill to make people skinnier, is compelling. There have, in fact, been a handful of biotech success stories, ones that have brought riches, fame, and other satisfactions to those involved. For some people there is an intellectual drive, a passion for learning, for uncovering the mysteries of life, or for fulfilling other unmet intellectual needs. And probably the greatest draw is the prospect of doing harm to humankind's universal nemeses, disease and illness. When Pandora of ancient Greek myth opened her box, pain, suffering, and disease came out—messages of terror sent by an angry god. A career in biotechnology, a donation to research, or even an investment in a risky biotech enterprise can be a response to that message. These motives continue to lure investors to biotechnology and alight the passions of scientists and other professionals in the business.

Hope springs eternal for biotechnology enthusiasts—and perhaps for good reason. Although it is still miniscule in proportion to the size of the industry, the rate at which the biotech industry is producing new drugs has been increasing. What had been a trickle of only one or two new products a year in the late 1980s has recently become a stream of between 10 and 20 a year. Profits from these products can be applied towards the development of more drugs. Furthermore, each experiment, from those that herald a new safe and effective drug to those that point towards failure, is a lesson learned. This knowledge can be applied towards the development of new drugs.

A hard look at successes and failures from both biotechnology and the more traditional approaches of medicinal chemistry reveals critical points in the research and development process. These critical points or bottlenecks helped give birth to the genomics industry and continue to nourish it today. To understand the role of the genomics industry, and how it is now contributing to the hopes of those in the drugmaking business, we must first consider some of the challenges facing drug developers.

Humankind's earliest drugs were most often plants whose medicinal properties were first found either by accident or through some form of experimentation. It is no surprise that plants contain pharmacologically active compounds, for ever since animals came into being, plants have had to endure their appetites. Plants, as well as fungal and animal species, have invented an enormous diversity of molecules that make themselves unpalatable by interfering with chemical processes in their would-be killers. Over the course of thousands of generations, animals developed molecules to detoxify the plant defenses, and a chemical arms race ensued. As an alternative to warfare, many species developed cooperative alliances, relationships that also have molecular correlates. Nature thus provides a rich source of molecules that specifically interfere or otherwise interact with human molecules or similar molecules in other animals.

Concurrent with the emergence of pharmacology, the science of drugs, came a reductionistic effort to extract and purify the active components of medicinal plants, fungi, and other natural materials. Efforts to modify plant-derived molecules were common soon after, followed by efforts to synthesize plantlike molecules from scratch. Finally, in recent years structural information on target molecules alone has become the basis for the synthesis of new classes of prospective drugs. The concept of a magic bullet, a synthetically made small molecule designed to hone in and either activate or deactivate a specific target, has become a reality. The protease inhibitors that are used to treat AIDS are good examples (see Chapter 13). Creating such drugs is challenging. A target molecule must be identified, candidate drugs (the bullets) must be formulated, the target molecule must be validated, tests must be conducted in animals to verify the intended effects and to discover any unintended effects, and finally, three stages of human testing (clinical trials) must be conducted to determine whether the drug is safe for intended users and more effective than existing treatments. Despite the advances brought by biotechnology, these steps still tend to be extremely costly and ponderously slow. It is not uncommon for drug developers to spend a half a billion dollars over the course of 10 years on developing a compound that fails to be worthy of regulatory approval.

Drug targets are molecules that act in disease processes. HIV protease and other essential HIV proteins are targets for AIDS drugs; molecules that relay nerve impulses meant to turn on acid production in stomach cells are targets for ulcer drugs; molecules that call artery-building cells into action at the site of a growing tumor are targets for anticancer medications; and molecules that direct artery-clogging fats into the bloodstream are targets for atherosclerosis medicine. Genomic companies sell tools that help identify targets in biological material. Gene chips and phage-display technology, which identifies protein to protein interactions, are examples of such tools. Some genomic companies apply these tools themselves and sell the resulting data. A list of 1000 tumor-specific proteins or a relational database of 20,000 human gene sequences present in the synovial fluid (located within the joints) of arthritis patients may be produced on a compact disk or on an Internet site, for example. Genomic companies may also sell software for analyzing and visualizing this information, such as a computer program that predicts and displays a three-dimensional image of any amino acid sequence or the functional motifs within any gene sequences. Or they may sell the results of such analyses, such as a set of 10 genes which are induced in the synovial fluid of all arthritis patients and that have sequence similarity to known inflammation-inducing genes.

There are several themes here. First, genomics companies deal in information. They sell biological data or data-producing tools. They intercept molecular messages, transduce them into electrical signals, capture them on computer systems, analyze them with computer programs, and communicate the results with pictures and words. Second, genomics companies capture, store, and analyze large amounts of biological information simultaneously. Tens of thousands of cells, proteins, and DNA or RNA sequences are considered at once.

Incyte and Human Genome Sciences were founded in the early 1990s, and since then hundreds of genomic companies have sprung up. Incyte, in particular, spearheaded the development of this new industry. Over 20 drugmakers bought access to Incyte databases, including 7 of the 10 biggest in terms of research spending. The deals were nonexclusive and created an instant revenue stream as well as rights to future royalties. Revenues steadily increased as Incyte added new partners and new database products. The company became the model for the industry, and when Incyte first generated profits, in fourth quarter of 1996, it fully distinguished itself from the legions of cash-burning life science research shops. Incyte did for biological information and genomics what Genentech did for the production of proteins and biotechnology. Incyte industrialized the acquisition of biological information, whereas Genentech industrialized recombinant DNA techniques. Each inspired a new industry.

Genomic companies apply innovative technologies to access, acquire, store, analyze, or display all sorts of molecular information (see Fig. 15.1). Some companies develop and use two-dimensional gel electrophoresis and mass spectroscopy to identify thousands of protein fragments. Others develop bioinformatic tools that identify functional motifs in gene sequences or ones that model molecular structures. Others create sophisticated detection devices for tracking the location of thousands of different proteins using thousands of different antibodies. Pharmaceutical and biotechnology firms have embraced genomics with uncommon vigor, formulating both in-house programs and multiple partnerships. As a consequence, a torrent of new drug targets has rained down upon the drugmakers. However, the role of genomics need not end there. Large-scale information-based approaches can be applied to additional portions of the drug development pipeline.

To validate a drug target, researchers must stimulate, inhibit, or otherwise perturb the molecule and then observe a desired consequence. Restoring the CTFR gene to cells derived from cystic fibrosis patients brings back desired ion conductance levels. Binding antibodies to the HER2 protein on the surface of breast cancer cells alters the cells' growth properties. Experiments such as these establish the target as a suitable starting point for drug development studies. Molecular perturbations can also be done on a large scale, and the outcome of each perturbation can be evaluated in terms of its effect on thousands of different molecules. For example, in order to learn more about the function of many different genes, several genomic companies are performing large-scale knockouts of yeast, fruit fly, worm, or mouse genes. Thousands of different genetically

Company	Market Capital (as of 11/00)	Business
Applied Biosystems (PEB)	$20.4B	Tools for sequencing, genotyping, etc.
Agilent Technologies (A)	$19.0B	Tools for gene expression, genotyping
Millennium Pharmaceuticals (MLNM)	$13.1B	Information and drug development
Human Genome Sciences (HGSI)	$8.8B	Information and drug development
Celera Genomics (CRA)	$3.8B	Information: sequences, protein interactions
Affymetrix (AFFX)	$3.7B	Tools for gene expression, genotyping
Incyte Genomics (INCY)	$2.0B	Information: sequences, gene expression

FIGURE 15.1 Leading genomics companies.

modified organisms are being created. The traits that each displays can be captured electronically along with the identity of the deleted gene. This is just one of many lines of approach in the quest for gene function, the subsection of the industry known as *functional genomics.*

The process of generating drug candidates has undergone a revolutionary change that parallels that of the genomics revolution. Rather than making a half dozen modifications on a single plant-derived toxin, thousands or even millions of modifications can now be made on any one of a growing repertoire of synthetic compounds. Combinatorial chemistry and other innovations in chemical synthesis boost the number of drug candidates to levels that were previously unheard of. Robots and computers track the compounds; their effects on drug targets are screened in a high-throughput fashion; and, once again, the data points go straight into a computer database. In this way a million different synthetically-made compounds can be tested to determine which bind most tightly to the protease proteins of HIV, for instance.

Before marching on to the next front of the genomics revolution, consider some of the unique challenges of the medical sciences. The ultimate subjects of pharmacological studies, sick human beings, are almost universally protected from a broad range of experiments. Generally, a drug cannot be administered to human beings until a large body of evidence suggests that it is likely to be safe and beneficial, or its use is deemed compassionate, a last resort for someone believed to be facing imminent death or endless suffering. When human experimentation is allowed, the scope of the experiments, known as *clinical trials*, is severely restricted. Pharmacology is thus disempowered relative to, say, geology, where minerals are vaporized and attempts to drill into the core of mother earth are tolerated. Though it may have noble aims, if the gods of Mt. Olympus were represented by the Sciences, then Pharmacology might appear among the lesser gods.

A large component of pharmacological research revolves around compensatory maneuvers, specifically the rather tricky task of modeling human physiology with laboratory animals and cultured cell lines. The Holy Grail here is a system that faithfully mimics the biology of sick people. Ideally such a model system will utilize the same molecular pathways as humans, display a comparable phenotype, and be predictive of events in humans.

Of course, the ideal has not been reached. The litany of cancer drugs that work in mice, but are useless (or harmful) in humans is testament to the challenge. A prospective drug may have unintended effects by acting on nontarget molecules, or it may be metabolized in unanticipated ways. Perturbations in target molecules may also yield unanticipated consequences. Nonetheless, animals are generally the best available human surrogates, and animal studies receive the lion's share of preclinical drug research and development funds. Sorting out just some of a drug's complex actions may require a decade or more of costly laboratory research. Hence, there is an economic pressure to both hasten the flow through this bottleneck in the drug development pipeline and to improve its filtering ability. (Couple the cost of these efforts with those of marketing and sales, and it is easy to see how the massive infrastructure of the highly-capitalized international drug firms tend to outleverage the scientific know-how and brilliant new ideas of the biotech boutiques.)

The drive for better disease models is itself a scientific and technological frontier, again one where anarchy may unceremoniously shine. To create better models for testing prospective drugs, researchers humanize other species by introducing human genes and tissues into them. Discarded human cells and tissues are used for drug testing. The rights of animals, dying people, dead people with living cells, the yet to be born, and others are openly debated. Headless human bodies are contemplated, as are embryo-derived tissues. Outrage is expressed, and scientific, moral, and religious issues are heatedly debated. Some ideas move forward, while others are suspended. Authority may spring forth spontaneously and may be applied when and where it is needed. Later, legislative bodies may establish rules, guidelines, or laws that reflect this authority. But this concerns xenobiology, the introduction of material from one species into another, and stem cell research, the study of cells that can develop into various portions of the body; what about genomics?

Genomewide information-based techniques can help in developing better experimental models of human disease and in evaluating drug treatments with these models. For example, virtually all new drugs currently are tested for toxicity in rats and monkeys or other primates. The status of these animals is evaluated after exposure to various doses of the drug. The liver is of particular interest because most foreign substances that enter the bloodstream are metabolized through the action of liver cells. However, the diet, lifestyle, and evolutionary history of these animals differ from those of humans, and so their livers do not function in exactly the same manner as a human liver. Animal toxicology studies can also be costly and time consuming. With microarrays and associated bioinformatic analyses, gene expression levels can be measured in immortalized human liver cells or other cell lines; patterns of gene expression changes that correlate with toxicity in humans may be found; and new models for predicting toxicity in humans can be evaluated. Cell-culture-based models may be found that are cheaper, better, and faster than existing whole-animal-based toxicology models, at least for screening large numbers of new drug candidates.

Drug treatments usually bring about side effects, unintended and often harmful responses to the drug. Molecular networks are complex, and perturbations (drug treatments) can propagate in multiple directions. A promising new antidepressant drug that acts on a particular target in the brain may also poison kidney cells. A potent killer of tumor cells may be highly toxic to the intestine and heart. Each of these responses has a characteristic gene expression profile in its respective animal or cell culture models. Thousands of new variants of these drugs can then be screened for the desired combination of gene expression profiles. This type of analysis, in which drug actions are evaluated in terms of the genomewide changes, is known as *pharmacogenomics* (not to be confused with *pharmacogenetics*, in which the influence of gene variants on drug actions are evaluated) and represents an enormous new market for genomic companies.

The genomics industry is also poised to serve physicians who conduct human clinical trials and the medical scientists and pharmaceutical companies that support them. Patients receiving medication are usually evaluated in many different ways. Blood tests, for instance, are routine. The pattern of molecules present in a patient's blood may speak volumes about the patient's response to treatment. A snapshot of gene expression in blood cells or biopsied tissue may say even more and could complement other monitoring practices. In their "Star Trek" application, the one which viewers of the popular science fiction television show of the 1980s immediately hit upon, gene chips and analysis devices are used to monitor patient health at the point of care, in real time and with remarkable precision. (Dr. "Bones" McCoy's medical scanners were also completely nonintrusive. Perhaps, someday . . .)

Another important commercial application of genomics is in segmenting patient populations by the gene variants that they possess (their genotype) and tailoring drugs to each group (i.e., pharmacogenetics). Many prospective drugs have failed to become commercialized not because they do not significantly help patients, but because they harm a segment of the patient population. For example, in 1926 the drug pamaquine was introduced as a treatment for malaria.[5] The drug helped most patients by killing infected blood cells, but a small percentage of patients developed severe jaundice and massive red blood cell loss, and then died. Pamaquine had contributed to their demise. The drug was discontinued. Thirty years later it was determined that patients vulnerable to pamaquine had a particular variant of the 6-phosphate dehydrogenase gene which made their red blood cells highly sensitive to the drug. In theory, patients could be screened for this variant, given pamaquine if they do not have it, and given another medication if they do. Better malaria drugs had been developed in the interim and pamaquine is no longer used, but the idea of stratifying patient populations by genotype has gradually taken root.

Genomic approaches enable the discovery of many more drug response-specific gene variants. Specific gene variants (and perhaps gene variant combinations) may be linked to drug responses, both positive and negative, by the same techniques that link them to genetic diseases. In these cases, finding a variant that affects only 2 percent of the patient population may be greatly advantageous to the other 98 percent, for if these 2 percent can be excluded from consideration, then a drug candidate that benefits 98 percent of the patients can move forward towards regulatory approval and medical use. In addition, microarrays and related technology may soon enable doctors to screen for many such variants quickly, cheaply, and accurately. Microarrays are already commercially available that detect p53 and HER2, gene variants that have been shown to influence patient responses to various cancer treatments. Through pharmacogenetics, drug companies will soon be better able to tailor "the right drug for the right person," and physicians will soon be better able to determine which drug is appropriate for their patient.

From target discovery through clinical trials and beyond, the genomics industry is beginning to meet the pharmaceutical industry's need for "better, faster, cheaper" means of developing new drugs. Genomic companies have been able to generate product revenues in a much shorter time frame than their biotechnology cousins, and they appear to have created a significant niche for themselves in the drugmaking business. This niche could grow significantly as information-based approaches spread through the drug development pipeline.

Genomic companies have largely retained the rights to their enabling technology. If the incentive to innovate and commercialize is preserved, then genomic companies will stay at the cutting edge of technology, the genomics industry will continue to grow rapidly and reinvest in new research and technology, and the business of making drugs will be profoundly altered. What about governments and private foundations? Might the genomics industry be derailed by large-scale efforts to put genomic information into the public domain? This is unlikely. Sequencing, microarrays, data analysis software, and other technologies available in the public domain are not keeping up with those available through genomic companies. The genomics industry may soon serve university research labs in much the same way that commercial computers and software serve university computer science departments. They may provide the tools for more advanced research. Furthermore, although publicly available information continues to be of great use to drug developers, it is unlikely that governments and private foundations will cater to their interests as genomics companies have, particularly for the later stages of drug development.

Could other factors derail the genomics industry? Certainly. Should legislative bodies, judges, or juries deny people of the rights to the biological information or technologies that they discover or produce, then the genomics industry could be severely hindered. On the other hand, should the ability to make a new drug depend upon scores of different proprietary genes and technologies, each owned by a different royalty-demanding company (the "tragedy of the anticommons," see Appendix), then the genomics industry may stagnate or become entrenched in litigation. However, this seems unlikely, for high-tech industries tend to be very dynamic and they find ways of working things out, particularly when companies' survival and people's incomes and investments are at stake. Business school graduates and successful businessfolk know artful ways of getting along. Dealmaking, cross-licensing, mergers, and acquisitions often obliterate further transaction costs, thereby lubricating industry's powerful engines.

Large-scale, high-throughput sequencing operations initiated revolutionary changes in the pharmaceutical industry. Rather than considering molecules one at a time, thousands upon thousands of molecules are now sampled in an instant. Many innovative techniques have sped up research. The polymerase chain reaction (PCR), for example, is used to rapidly generate millions of copies of DNA. However, genomic technologies are beasts of a different nature. Genomic technologies not only speed up research tasks, they forge a more intimate relationship with the subject of study. A more intimate relationship!? From a factory of automated machines and computers!? Yes, for two-dimensional protein gels, expressed sequence tags (ESTs), gene expression microarrays, and subsequent genomic technologies allow us to eavesdrop on molecular messages within us—and not just isolated messages, but the whole gamut of messages. Imagine huge alien creatures tapping our phone lines, intercepting our satellite signals, and peering into our homes. They could become quite familiar with us.

As we learn to interpret the molecular messages within us and figure out how to formulate and direct our own molecular messages, we once again gain new powers over our destiny—and are burdened with new responsibilities. For example, the ability to segment populations according to genotype and to tailor drugs to particular genotypes will create more choices for drugmakers. Which segments of a patient population will they focus more of their resources on? Industry is industry. Companies seek to gain market share, increase stockholder value, and so on. They are subject to the law of economic selection. Power and responsibility remain with people, and so the genomic industry, as well as the biotech, pharmaceutical, and health care industries, will continue to be greatly influenced by the incentives we create through our health care choices, the economic and legal systems we create, and our allocation of tax revenues and other collective resources. Jeremy Rifkin, president of the Foundation on Economic Trends and perhaps the most prominent genomic industry watchdog, posed this issue in the following way:

> It needs to be stressed that it's not a matter of saying yes or no to the use of technology itself . . . Rather, the question is what kind of biotechnologies will we choose in the coming Biotech Century? Will we use our new insights into the workings of plant and animal genomes to create genetically engineered "super crops" and transgenic animals, or new techniques for advancing ecological agriculture and more humane animal husbandry practices? Will we use the information we're collecting on the human genome to alter our genetic makeup or to pursue new sophisticated health prevention practices?[6]

These are questions for all of us to consider as we shape the future of our society.

The pharmaceutical industry is a key driving force behind the genomics industry, and the pharmaceutical industry is in turn driven by a deep and widespread desire to end or mitigate suffering. However, as many have noted, pills may not provide the best solution to many of our ills. Good health can often be achieved in other ways. Preventative measures, such as changes in diet and exercise, may be more beneficial than any pill. Genomics and the genomic industry could also be directed towards evaluating responses to lifestyle changes, herbal remedies, and the like. Little has been done in this area, but if proper incentives are created, then we all may benefit enormously from such studies.

16

The SNP Race

Biomedicine's unusual blending of public and private funding and commercial and noncommercial interests was again demonstrated in April of 1999, when a group of 10 pharmaceutical companies and one charitable foundation announced the creation of a non-profit enterprise known as the SNP Consortium. The consortium's goal was to identify and map 300,000 common human DNA sequence variations (known as *single nucleotide polymorphisms* or SNPs) within two years. As with the Merck/Washington University ESTs, these chromosomal map points would be made available to all researchers and would not be patented. Oddly, though former Merck executive Alan Williamson came up with the idea for this consortium, Merck chose not to join. The 10 drugmakers that did participate contributed a total of $30 million to the effort, while the Wellcome Trust donated $14 million.

The SNP Consortium would complement a U.S. government–funded project to create a database of between 60,000 and 160,000 SNPs that was announced in December of 1997.[1] The publicly available database, known as dbSNP, began in the fall of 1998 and had grown to over 15,000 entries by the summer of 1999.[2,3] Francis Collins, a highly respected molecular biologist and head of the HGP, crafted that effort, and one of his intentions was to avoid a situation in which medical researchers would be "ensnared in a mesh of patents and licenses"[4] (the anticommons dilemma described in the Appendix). Collins had drummed up $30 million for the project from an unprecedented consortium of 18 NIH agencies. Apparently it was not enough, for two years after the government-backed effort was initiated a significant number of pharmaceutical companies were still concerned about private enterprises owning the rights to a significant number of SNPs. One of their missions in starting the SNP Consortium was to prevent researchers from being "held hostage to commercial databases."[5]

Predating both public initiatives, in July of 1997 the French genomics company Genset announced that it would develop a 60,000-SNP map of the human genome. Genset's mapping project was funded through a $42.5 million partnership with Abbott Laboratories, a pharmaceutical company that did not join the SNP Consortium (see Fig. 16.1).

Celera officials had declared their intention of selling SNP data at the company's start in 1998. Incyte also intended to commercialize SNP information, as did Curagen Corporation, whose November 1998 announcement of the discovery of 60,000 SNPs led to a brief 50 percent rise in its stock price. At the time of the SNP Consortium announcement, SNP mania was growing rapidly. Could the noncommercial outfits outwit the patent seekers? Could the drugmakers outwit the pesky genomic upstarts through an alliance among themselves and Wellcome, the world's largest charity?

Whereas the Celera genome-sequencing announcement of 1998 put the public genome project on the defensive, the SNP Consortium's 1999 announcement set the private projects on edge. Genomic company stocks took a beating. Nonetheless, genomics industry officials took the high road and generously insisted that the publicly available SNPs would simply complement their efforts.[6] Industry investment analysts were less kind. Several questioned the consortium's ability to produce the 300,000-SNP map in a timely manner. Robert Olan of the investment bank Hambrecht & Quist went so far as to declare that "there has been no academic lead venture that has ever been competitive with the commercial operations."[7]

The ability to readily detect human sequence variations and link them, either directly or indirectly, to human traits, diseases, or propensities is a frightening power, for its misuse could have grave consequences. Groups of people with particular sequence variations could be targeted for harm. Others could be unjustly favored. It is no surprise that the SNP race, an offspring of the race to sequence a composite human genome, is fraught with as much

Institution	Date	Goal	Funds (if known)	Strategy	Status (if known)
Genset/Abbott	Jul 1997	60K SNPs for mapping	$42.5M	Mapping disease genes, initially	
Celera	May 1998	Catalogue of human variation		Via shotgun sequencing alignments	2.4 million unique SNPs as of 9/2000
Incyte/Hexagen	Aug 1998	100K SNPs in cDNA		Mostly by sequence alignments	>70K as of 8/2000
NHGRI/NCBI	Sep 1998	60–100K SNPs for dbSNP	$30M	Submissions from throughout the scientific community	26.5K submissions as of 3/2000
Curagen	Nov 1998	60K SNPs in cDNA		Integrated genomics based drug discovery and development	>120K cSNPs as of 4/2000
SNP Consortium	Apr 1999	300K SNPs evenly distributed throughout the genome by 2002	$44M initially	To make SNP data available to the public without intellectual property restrictions	296,990 mapped SNPs as of 8/2000
Various ministries of Japan	May 1999	100K–200K SNPs in coding region of genes by 2001		Create public database from SNPs found in 50 Japanese individuals	
HGS/Compugen	Mar 2000	500K SNPs in cDNA		Via EST and genomic DNA alignments	Plan to be done by 3/01

FIGURE 16.1 Large-scale SNP discovery initiatives.

controversy as is its madcap parent. The focus here is on the underlying science of SNPs and the developments that are bringing them into the forefront of public debate. At the time of this writing the strength and scope of SNP patent claims are uncertain. The ability to keep personal SNP information private is also unclear.

SNP stands for *single* *nucleotide* *polymorphism*. A single nucleotide polymorphism may be, for example, a site on the genome where in 31 percent of the human genomes there is a T, whereas in the other 69 percent there is a C. Single nucleotide polymorphisms are the most common type of minor sequence variation between individuals, accounting for approximately 85 percent of the sequence variations. SNP, as the term is used today, also refers to other minor sequence variations, such as those regions of the genome where some individuals have a few more or a few less nucleotides than others do. For a site on the genome to be considered a SNP, greater than 1 percent of the entire human population must have a variation at that site. If the variation occurs in less than 1 percent of the population, than it is somewhat arbitrarily designated as a mutation.

SNPs are old. They probably began as mutations and spread through the population over the course of thousands or millions of generations. Some sets of SNPs are so old that they occur in both apes and humans. These SNPs probably existed prior to speciation, when the two species evolved from a single ancestral species.

There are two main types of SNPs, (1) those that are likely to affect traits, and (2) those that are not. The vast majority of sequence variations are in portions of the genome that do not encode genes. By definition, these SNPs do not affect traits. The majority of human genome does not encode genes, and because there is usually little or no selective pressure against them, mutations that occur in these sequences may spread through the population. Other SNPs occur within either a gene's regulatory or protein coding sequences. Some of these SNPs make no discernible difference in protein composition or expression, whereas others create changes.

Estimates of the frequency of SNPs in the human genome range from one SNP per 350 bp of genomic sequence to one SNP per 1000 bp.[8] These studies, which are based on sequencing a portion of genomic DNA from a diverse set of individuals, indicate that the human genome may harbor as many as nine million SNPs. In the coding sequence of genes there are about 400,000 SNPs, of which approximately one-half are predicted to alter the amino acid sequence of an encoded protein.[9,10] These 200,000 or so SNPs are likely to affect variation in human traits; each variation is responsible for a unique gene allele. SNPs that vary protein sequences occur less frequently than SNPs that do not alter protein sequences, presumably because natural selection has acted against such variations.

An undetermined number of SNPs occur in gene regulatory sequences and are also likely to contribute to human variation. These SNPs often lie upstream of the protein-coding region where they influence how much protein is made, and where, when, and under what circumstances it is made. SNPs that influence protein composition or expression (and are thereby likely to affect traits) are known as *functional* SNPs. All others are *nonfunctional* SNPs.[11] The Holy Grail in the SNP race is the identification of all functional SNPs and the deciphering of the connection between these DNA sequence variations and human diseases, disease propensities, and treatment responses.

So, a few hundred thousand SNPs may be responsible for nearly all heritable human variation. What about the millions of other SNPs? Are the nonfunctional SNPs worthless? Subsets of these SNPs have guided researchers to numerous disease-causing gene sequence variations (mutations). Clearly, they are not worthless. (The use of sequence variations in identifying disease genes was described in Chapters 11 and 16.) Newly identified SNPs act as additional markers on chromosome maps. Denser maps often make it easier to hone in on disease genes and on functional SNPs. So, nonfunctional SNPs are highly valued by those trying to map genes. They may also be of great value to those studying human evolution.

Many of the most common heritable diseases, however, defy positional cloning. No matter how good our gene maps are, it is not always possible to locate a gene for a particular disease. And even when a correlative gene is found, most often it cannot account for the disease pathway, or it is not implicated in most cases. Thus, as we saw in Chapter 12, an understanding of the most common genetic diseases has remained out of reach, despite a flurry of disease-gene discoveries. Headway is stalled by these and other complexities inherent to the disease processes.

SNPs may offer a way around these difficulties. Many scientists now believe that common variants (SNPs) make significant contributions to the development of common diseases, such as arthritis, Alzheimer's disease, and diabetes.[12] Naturally, they are eager to further develop and test this hypothesis, and their interest is driving rapid advancements in SNP detection and analysis techniques.

At the forefront of genetic research on common, yet complex, diseases is a simple mathematical method known as *association analysis*. Neil Risch of Stanford and Kathleen Merikangas of Yale University have argued that the "method that has been used successfully (linkage analysis) to find major genes has limited power to detect genes of modest effect, but that a different approach (association studies) that utilizes candidate genes has far greater power, even if one needs to test every gene in the genome."[13] An association study may work as follows: DNA is obtained from a set of affected individuals and their parents. For all genes that are heterozygous (having two alleles) in a parent there is a 50 percent chance of transmitting any one particular allele to any one child. However, if an allele contributes to the affected phenotype (i.e., is dominant), then this allele would be transmitted to an affected child in much greater than 50 percent of the cases. This measure of the association between alleles and traits may allow researchers to more rapidly hone in on alleles that contribute to disease processes.

Thaddeus Dryja and colleagues at Harvard University used association studies to implicate several distinct alleles that cause hereditary blindness.[14] Using samples derived from hundreds of patients with retinitis pigmentosa and other sight disorders, Dryja and colleagues looked for variations in 14 genes known to function in the specialized cells of the retina. They found defects in 7 genes that were very strongly associated with the disorders. These findings quickly led to new diagnostic powers and new insights into hereditary blindness. Because their studies were probabalistic in nature, they did not require knowledge of the map position of genes, nor did they require that the genes be traced through the lineage of large extended families with many affected members.

Common alleles are characterized by SNPs. Alleles often differ at one SNP. The full power of association studies will come with the application of new technologies that allow the identification of SNPs in thousands of individuals. Dryja and colleagues applied foreknowledge of 14 genes to their studies, but with new technology prior knowledge of gene function may no longer be necessary. If thousands of SNPs could be assayed simultaneously, then associations could be mined from the resulting data set. Like the sequence databases described earlier, a SNP/disease database would be cumulative, and as the database grows larger the power to find statistically relevant associations will increase.

Furthermore, not only may single-gene disease associations be readily found through such a database, but the more elusive combinatorial gene actions may also emerge. A computer could rapidly test all combinations and calculate statistically derived measures of confidence. People with allele A of gene 1 may have just a twofold greater chance of getting diabetes before a particular age, while people with allele B of gene 2 may have a 1.5-fold greater chance, and those with allele C of gene 3 may have only a 1.3-fold higher chance. However, the people with all three alleles may have a greater than 90 percent chance of becoming diabetic. If this were the case, then this knowledge would stimulate researchers to probe the mechanism by which these three alleles interact in bringing about diabetes. New diagnostic powers and new treatment leads would result.

Association studies need not be limited to genetic inputs. Environmental conditions, dietary habits, drug treatments, and other information could also be fed into a computer database. The next chapter will discuss how new discoveries and new health care knowledge may be derived from such databases. Now let's consider how such databases are being built.

There are two main avenues towards SNP discovery. In both cases, one starts with DNA derived from individuals of diverse heritage. One may then sequence a particular region of the genome in each of these DNA samples. SNPs become obvious when one aligns the sequences (see Fig. 16.2). Alternatively, one may take advantage of the hybridization properties of DNA. Single-stranded DNA molecules will anneal with each other. However, anomalies in the structure of the DNA will occur in places where there is a sequence variation. These anomalies can be detected using a variety of different techniques, including gel electrophoresis and mass spectroscopy.

Sequencing may be accomplished using automated sequencing machines, and SNP discoveries may be a by-product of sequencing projects. In EST factories, for example, commonly expressed genes are sequenced many times over. Computer programs have been created that align the sequences, carefully con-

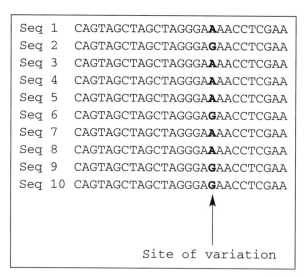

FIGURE 16.2 Sequence variations (SNPs) can be discovered by aligning sequences from a variety of individuals. Slightly more than one SNP is found for every 1000 base pairs of human DNA sequence examined.

sider sequence quality, and determine which variations are due to sequencing errors and which are due to SNPs.[15] SNPs derived from ESTs are particularly valuable because they come from expressed genes and are therefore more likely to be functional SNPs than those derived from randomly selected genomic DNA.

A more efficient way to sequence a known region of the genome is through a chip-based sequencing-by-hybridization method (first introduced in Chapter 14). Sequencing chips have been made in which entire genes are represented by a tiled array of short overlapping fragments.[16] The way these SNP-detecting chips work is as follows. To survey a region of 1000 base pairs, one begins with 1000 DNA sequences, each 25 base pairs long, where each sequence overlaps the next in all but one nucleotide. Each of these 1000 sequences is duplicated three times in such a way that the nucleotide in the center of the sequence is changed. Thus each of the four nucleotides (A, T, C, G) is represented in the center position in one of the copies. Because each of the original sequences varied by only one nucleotide, the center position on each of them represents a different site, and taken together, all of the sites are represented. All 4000 sequences are then synthesized or spotted at discrete positions on a glass chip. This chip is then hybridized to a fluorescently labeled DNA sample. Only exact matches light up, and the pattern of spots reveals the precise sequence of the selected region of the genome. The sets of four arrayed DNA sequences with variations at each site along the sequence enable the detection of all possible sequence variations (SNPs). For each set of four arrayed sequences, only one will light up, unless the sample is from a heterozygous individual (one who carries two alleles of the gene), in which case two out of the set of four fragments should light up.

In a report published in the journal *Science* in 1998, David Wang and colleagues at the Whitehead Institute at the Massachusetts Institute of Technology and Affymetrix described their use of sequencing chips to search for SNPs along two million base pairs of genomic DNA.[17] Using DNA samples from 23 individuals, they uncovered 2748 likely SNPs.

A variety of other SNP-detecting techniques utilize the differential hybridization properties of SNP containing DNA, but do not bother sequencing the entire fragment in each individual. A SNP containing fragment may be detected by its altered mobility in a gel or by its differential response to certain DNA-cleaving enzymes, for example. New SNP discovery techniques continue to emerge as the SNP race goes on.

To do association studies and thereby link SNPs to human diseases, disease propensities, and treatment responses, it is not sufficient to simply identify SNPs in the collective human genome. One must also be able to determine an individual's set of SNPs (their *genotype*). Association studies may require that tens of thousands of individuals be genotyped at thousands, or even hundreds of thousands, of different sites on the genome (*loci*).[18] Furthermore, if drugs are to be tailored to an individual's genotype, one must have highly accurate and cost-effective means for SNP detection in the clinic. Sequencing the entire human genome of each patient is impractical, so "faster, better, and cheaper" means of assaying SNPs must be found.

In their landmark paper, Wang and coauthors showcased a prototype SNP chip that is capable of simultaneously genotyping 558 loci.[19] The SNP chip is an adaptation of the sequencing chips just described. The difference is that chip real estate that had previously been used to sequence invariant regions of the genome has been replaced with DNA fragments that only hybridize to sequences surrounding newly identified SNPs. Whereas, 149 different types of chips were previously needed to identify 2748 SNPs, with this method only one chip design was necessary to genotype 558 SNPs.

SNP discovery and devices for high-throughput multi-SNP genotyping are purely technology/engineering challenges. It is the stuff of Moore's Law, of innovations, miniaturization, commercialization, patents, and profits. Taxpayers and foundations act as co-conspirators. And the result? Technological advances, such as SNP chips, are coming at a furious pace. SNP databases are rapidly expanding. There is little doubt that in short order nearly all human SNPs will be discovered, and that cheap and reliable means for detecting SNPs will become available.

Will these research tools be widely available? This was the concern that helped put first 18 NIH agencies and then 10 pharmaceutical companies and the Wellcome Trust into high gear.

With these powerful entities at work, the answer is assuredly yes! Researchers will have access to SNPs and SNP detection technology (and gene sequences too). SNPs are flowing rapidly into public databases, free for all to use. Patent owners will dare not intervene, that is, unless someone makes a commercial product from a contested SNP (or gene). And then profiteers will do as they do. They will negotiate a mutually satisfying arrangement. Otherwise they will be threatened with loss of income. There is no incentive for holding medical progress hostage. And SNP detection technology? A free market, high demand, and vibrant competition, the economic engines of our economy, will assure that SNP chips or alternative technologies are affordable and available.

Demand for SNP information is premised on the expectation that researchers will be able to pin particular SNPs or sets of SNPs on specific diseases, disease susceptibilities, and treatment responses—the Holy Grail. It is hoped that this knowledge will lead to a deeper understanding of the molecular basis of diseases, enable drug developers to tailor medicines to particular genotypes, and allow physicians to better match patients and treatment choices. But is this, too, a certainty? Just how will SNPs and SNP technologies bring about a deeper understanding of human biology and better health? Theoretical studies do not always lead to practical benefits. In fact, the utility of association analyses using SNPs is hotly contested.[20-22] Furthermore, as was discussed in Chapter 12, the path from a disease-associated allele to a cure is usually unclear.

Of course, there is no script that outlines precisely how this plot will unfold. Although the Human Genome Project has a 15-year plan and each genomics enterprise has a long-term business plan, the specifics on how new medicines will emerge are necessarily absent. No one knows for sure. Capturing biological information in electronic form is a cinch. But, how will these genomic discoveries, the consequence of "faster, better, cheaper" technologies, be translated into longer, better, and richer lives?

Of course, we are writing our own script, and we amend it as we go forward. In the final chapters of this book let's further consider attempts to bootstrap our way into a new age of health care.

17

From Information to Knowledge

A *technological* revolution has provided researchers with an unprecedented level of access to genomic information. The sequences of nearly all human genes have been identified. Thousands of sequence variations (SNPs) can be assayed simultaneously. Researchers can determine the gene expression level of tens of thousands of different genes, from hundreds of different bodily tissues. Thousands of different proteins can be sampled from these same tissues, and many other large-scale assays can also be applied. The net result is a monumental pile of electronic data, a torrential flood of encoded numbers. An additional revolution may be required to distill knowledge from this sea of data, for transducing the genome is one thing, but eliciting new truths (deciphering the genome) is another, and eliciting better health is yet another. What scientific framework will meet these challenges and thereby change the world?

Traditionally, to test a hypothesis, one collects data and extracts the numbers that will either support or negate the hypothesis. To determine whether altering gene A causes the expression of genes B, C, and D to go up, one does the experiment, collects the data, and presents the results. The sea of biological information collected through genomic technologies compels a new approach. First, one collects a large amount of biological data, and then one generates and tests hypotheses. All genes may be altered one by one, for example, and all genes may be assayed each time. In this case, initially the only hypothesis may be that something interesting will be found within the resulting set of data.

This database discovery approach can be seen in the work that John Taylor, Leigh Anderson, and colleagues conducted at Argonne National Laboratory in Illinois in the early 1980s, the predawn of the genomics era. Recall that the Argonne group had identified 285 spots on X-ray film, each representing a distinct protein expressed in each of five cancer cell types. The intensity of these spots reflected the relative abundance of the underlying proteins. Multivariate statistics was used to classify the cell types; principal components analysis and cluster analysis reduced the dimensionality and enabled the relationships between the cell types to be visualized. Yet another mathematical technique, *discriminant analysis*, could have been used to rank the 285 proteins according to how useful they are in discriminating between cell types. In each case, mathematical analyses and computation take a central role.

Principal components analysis is somewhat like projecting the shadow of a three-dimensional object onto a two-dimensional surface. There are infinitely many ways of doing this projection; principal components analysis chooses the way that retains the greatest amount of information. Many of the 285 proteins were nearly invariant from cell type to cell type. Principal components analysis dismisses these uninformative dimensions, combines many of the remaining dimensions, and allows one to visualize the cell types as distinct points in two- or three-dimensional space.

To get an idea of how cluster analysis works, imagine protein expression data displayed on a spreadsheet where cell lines are listed as rows, proteins are listed as columns, and protein expression levels are placed in the matrix created by the intersection of the rows and columns (see Fig. 17.1). A measurement of similarity between pairs of rows is provided by a function that takes into account the differences between each pair of the protein expression values.[1] If the differences in corresponding protein expression values are slight, then the function yields a low value. If the cumulative differences are large, then the function yields a high value. Next, pairs of rows that are similar to each other are grouped together into clusters. A similarity tree, also known as a *dendrogram*, can be used to display the relationship between the cell types (see Fig. 17.2). If the experimental system is robust, then replicas samples from the same cell type will cluster tightly together. The Argonne group attained this measure of reproducibility.

	Sample 1	Sample 2	*Sample 3*	Sample 4	Sample 5	*Sample 6*
<u>Gene 1</u>	High	Low	*High*	High	High	*High*
<u>Gene 2</u>	High	Low	*High*	Low	High	*High*
<u>Gene 3</u>	**Low**	**Low**	***High***	**High**	**Low**	***High***
<u>Gene 4</u>	High	High	*Low*	High	High	*Low*
<u>Gene 5</u>	**Low**	**Low**	***High***	**High**	**Low**	***High***
<u>Gene 6</u>	**Low**	**Low**	***High***	**High**	**Low**	***High***
<u>Gene 7</u>	High	High	*High*	High	Low	*High*

FIGURE 17.1 Information from biological samples can be displayed on a spreadsheet. In this simplified version the expression values of seven genes in each of six biological samples are displayed. Genes and samples can be independently clustered together based on the pattern of expression values. A gene cluster is shown in bold. A sample cluster is shown by italics.

The application of such mathematical analyses to genomic information may be of tremendous medical value. To help understand how useful it may be, consider the ubiquitous and enduring blood and urine tests.

Doctor visits frequently require blood and urine samples. Laboratory tests are conducted to determine the level of a few dozen specific molecules (*analytes*) in these bodily fluids. Within a population of healthy individuals, each blood or urine analyte exists within a unique concentration range. Particular analytes, or, more often, combinations of analytes that fall outside their normal ranges, signal particular health problems. Therefore, doctors routinely use these tests to help make diagnoses and monitor patient progress.

There are thousands of different molecules within blood and urine, and it took many decades of research to identify and characterize those of diagnostic value. Molecules were (and still are) often tested one by one, usually after being functionally linked to a disease process. In March of 2000 the *New England Journal of Medicine* published a report on the discovery of one such molecule, C-reactive protein, shown by Harvard Medical School researchers to be of value in assessing heart attack risk.[2] Eventually,

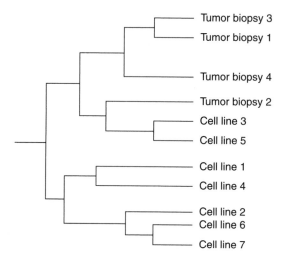

FIGURE 17.2 A dendrogram displays the relationship between biological samples based on their pattern of gene expression. Cultured cell lines numbered 3 and 5 are a better match to the tumor tissue than are other cell lines. Based on the author's actual data.

researchers determined the concentration range of numerous analytes that are associated with good health and the concentrations that are associated with various medical problems. This knowledge emerges slowly and sporadically.

In healthy individuals, proteins exist within particular tissue-specific concentration ranges, and combinations of proteins that fall outside their normal range are likely to be indicative of particular disease states, disease propensities, or treatment outcomes. The Argonne group pioneered a means of tapping into this information. Their methodology allowed the relationship between cell types to be instantly solved. The cells could be characterized by their protein expression pattern. The information already existed in the database, and it was only a matter of applying the appropriate mathematical tools. Then—poof!—like a movie in which a chemist stands beside a small laboratory explosion, the experiment culminates and one cries "Eureka!" Except that here it was done silently and electronically.

Protein expression databases had a slow start. Initially, two-dimensional gel electrophoresis systems performed inconsistently. Many proteins could not be separated, and it was often difficult to identify proteins for further analyses. Most disappointingly, few scientists caught on to the idea of information-based analyses. It was still on the fringes of molecular biology research.

Interestingly, those on the fringes tend to cross paths. The same analysis techniques that the Argonne group performed on protein expression values can also be applied to mRNA expression data, and in the late 1980s Taylor and Anderson collaborated with Leonard Augenlicht, the early microarray pioneer. However, initially this collaboration resulted in a only a few rather low-profile publications.[3] Following a course somewhat analogous to that of protein analysis, information-based gene expression analyses did not mature until the late 1990s. Anderson's protein data and Augenlicht's mRNA data were sufficient for crude studies. More advanced technologies and larger sets of data would come later.

A large, high-quality data set is a prerequisite for any information-based analysis. GenBank and other large-sequence databases facilitated countless studies. As the sequence databases grew, the bioinformatic toolbox expanded and powerful new analyses emerged. High-density microarrays (or gene chips) for capturing gene expression information became widely available in the late 1990s. A microarray that monitored virtually all yeast genes was created.[4] Other microarrays monitored the expression of tens of thousands of human genes. These technology advances in turn spurred the development of many other innovative information-based analyses. By the year 2000, scientific journals were featuring such reports on a weekly basis.

Cancer specialists were among those most interested in microarray technology. Tremendous efforts have been made to better characterize different types of cancers, so as to enable treatments to be tailored to each type. Some classification tests assay one or more of the gene sequence variations that underlie the cancer phenotype. Other tests examine the morphology (physical shape) of the cancer cells. The global gene expression assay is yet another tool in the classification toolbox, and what a powerful tool it promises to be!

Consider acute leukemia. These pernicious cancers of the blood have been classified into two groups, *acute lymphoblastic leukemias* (ALLs) and *acute myeloid leukemias* (AMLs). ALLs are derived from lymphoblast cells, while AMLs are derived from myeloid cells. ALLs are treated with corticosteroids, vincristine, methotrexate, and L-asparaginase, while AMLs are generally treated with daunorubicin and cytarabine. Classification by conventional means relies on a series of expensive, time-consuming tests. A team of researchers at Massachusetts Institute of Technology's Whitehead Institute devised a way to classify acute leukemias using microarray data.[5] Todd Golub, Donna Slonim, and coworkers analyzed a set of 259,046 gene expression values, 6817 genes in each of 38 patient samples (25 ALLs and 13 AMLs). Through mathematical analyses they found that the expression values from particular subsets of these genes could identify the leukemia type with 100 percent accuracy.

They then wondered, "If the AML-ALL distinction were not already known, could it have been discovered simply on the basis of gene expression?" The team asked a computer to align the samples in two groups using a clustering algorithm known as a *self-organizing map*. The self-organizing map looked for two points in 6817 dimensional space around which the data points appear to aggregate. The 38 samples were then partitioned by their distance to these two points. Of the 38 samples, 34 partitioned in accordance with their previous classification. Subsets of genes were tested for their ability to predict the two cell types. Those subsets that were the best predictors were used to run the self-organizing map algorithm a second time. The classification scheme improved, and all samples tested were now correctly partitioned. Thus, an iterative process was disclosed, one in which the samples (columns if you imagine the data on a 2-D matrix) are sorted, the genes (rows) ranked, and the columns are then resorted based on a subset of genes.

A group of researchers at Stanford and the U.S. National Cancer Institute provided further validation of the utility of gene expression microarray analyses in a report published in *Nature* magazine in January of 2000.[6] The subject of their study was diffuse large B-cell lymphoma (DLBCL), another deadly cancer of the blood. Approximately 1.8 million gene expression measurements were made using 128 microarrays and 96 normal and malignant blood cell samples. A clustering algorithm was used to group patient samples based on these gene expression values. The DLBCL samples split into two distinct groups. These groups each included normal B cells at distinct stages in development. B cells continually develop so as to replenish the body's ability to mount immune responses. From the microarray analysis, it appeared as if cancer could arise from cells at distinct stages in development. This, alone, is an important finding, for it allows insight into the mechanism by which cancers develop. However, the Stanford/NCI study offered even greater rewards. A simple mathematical association study revealed that the two newly discovered DCBCL classes defined new prognostic categories. Patients with one subgroup of DCBCL had a much higher chance of survival than those of the other subgroup. There is no doubt that this information will now be used to tailor treatment plans to each of the two newly defined DCBCL types. With the application of a clustering algorithm, the Stanford/NCI group, which included Ash Alizadeh, Michael Eisen, and Patrick Brown of Stanford, Louis Staudt of the NCI, and 27 others, had not only classified cell types but had rigorously and mathematically defined two new diseases.

Analyses such as these blood cancer studies offer further validation to a scientific approach based on transduction of fundamental information from the molecules of life and the *in silico* derivation of hypotheses based on this information. It is not yet clear how widespread this approach will become, but it is clearly capturing the attention and igniting the imagination of many biomedical researchers. Pharmacologists, for example, may recognize that the techniques applied in the blood cancer studies may be very relevant to the development of new animal models for human diseases. Recall that drugs must be tested on animals or cultured cells and that researchers are constantly seeking better models for predicting both harmful and beneficial drug responses in humans. Imagine columns on the gene expression matrix representing distinct samples of diseased human tissue and additional columns representing possible new animal or cell culture models. A good model system may be expected to cluster tightly with diseased tissues. They may have similar gene expression profiles under a variety of conditions and may therefore mimic the drug response of the disease tissue. Pharmaceutical companies are hotly pursuing such research.

Researchers found that expression information can also be used to interrelate genes in a manner that is entirely analogous to the way in which gene sequences are used.[7] Gene sequence similarity values are derived from algorithms, such as BLAST, which measure how well gene sequences match each other, and the resulting information defines gene family membership and provides important clues to gene function. Clustering algorithms, like those used by the Argonne, Whitehead, and Stanford/NCI groups, can also be used to create gene families based on mRNA or protein expression patterns. In this case, genes are listed as rows on a spreadsheet, biological samples (liver, kidney, muscle, etc.) are listed as columns, and the matrix contains the corresponding gene expression values (mRNA or protein levels). Genes of similar function tend to cluster together.[8] Thus, clues to gene functions can be found via similarities in expression patterns, as well as through sequence similarities.[9] Again, here gene function hypotheses are derived from the mathematical analysis of digitized sets of data rather than from directed molecular studies conducted in the laboratory. It is also worth noting that the source of much gene expression information is DNA sequence-bound, existing in often cryptic regulatory regions upstream of protein coding sequences, and that links between sequence and gene expression databases are enabling the identification of such regulatory sequences. If sequence similarity upstream of known genes is correlated with common patterns of gene expression, then it is likely that these sequences regulate expression.

Clustering algorithms can also be applied to drug candidates. John Weinstein of the NCI has led an effort to relate the chemical structures of drug candidates to gene expression patterns derived from an 8000-gene microarray, tumor cell growth response profiles derived from a panel of 60 tumor cell cultures, and additional information.[10] By 2000 the NCI's Development Therapeutics Program had processed over 70,000 chemical compounds. Weinstein, Uwe Scherf, then of the NCI, and coworkers classified a small subset of these drug candidates, ones whose mechanism of action was at least partially known. They found that compounds that act on the same molecular targets produced very similar patterns of tumor cell growth inhibition and thus clustered together. For compounds that have multiple molecular targets the clustering data gave hints as to which mechanism of action was more relevant to a compound's ability to inhibit cell growth. Most important, this clustering tool provides a way to classify previously uncharacterized compounds into meaningful function-associated groups.

Weinstein and colleagues also clustered the same compounds based on a mathematical measure that related their activity across the 60 tumor cell cultures to the expression pattern of genes in the same cell cultures. Some clusters were nearly the same as in the cell growth study, while others differed markedly. Some clusters seemed to suggest meaningful distinctions between subclasses of compounds. Three compounds that inhibit an enzyme that unwinds replicating DNA clustered next to two compounds that break a particular class of molecular bonds. Both result in broken DNA bonds—suggesting that their similar gene expression profiles are the consequence of broken DNA bonds. What significance do these kind of findings have? "Our aim is exploratory," admitted Scherf and coauthors: "We obtain clues, generate hypotheses, and establish context rather than testing a particular biological hypothesis in the classical manner."[11] Dominant tumor-killing properties and toxicities of the three DNA unwinding inhibitors may be a consequence of their DNA bond-breaking actions.

Do explorations such as these represent tantalizing leads for the discovery of new truths or do they represent information overload, a dizzying array of tenuous half-truths, unconfirmed hints floundering amid a bedazzling spectacle of data? For these database explorations to be worthwhile they must lead to new truths, not just any truths, but truths deemed worthy of passing on to others, findings that will lead to new treatments or new disease prevention strategies. This is "knowledge," and if a significant amount of knowledge is obtained in this way, then this would signify a genuine *scientific* revolution.

Through genomics biomedical researchers are stumbling into uncharted territories. Very few are accustomed to manipulating massive data files in anything more than in the most rudimentary of ways. Fortunately, others have voyaged into these realms. Indeed, geologists, ecologists, astronomers, as well as marketing professionals, salespeople, advertisers, business analysts, investors, political campaign directors, and many others went electric long before most biologists. They may study mineral deposits, migratory birds, interstellar matter, consumers, products, stocks, or voters rather than genes and proteins. No matter! Bird populations, product sales, banking and investment activities, opinion polls, and so forth can be electronically monitored and stored in enormous relational databases. Computer algorithms are used to mine these databases for information that helps solve scientific riddles or determine which products to stock, which people to target, what type of promotions to run, etc. *Database mining* is the term for this work, and some of the same database mining algorithms that enable soda pop makers, for instance, to attract more customers could be applied to the pursuit of better disease treatments. And some of the same people who create computational tools for the soft drink manufacturer could redirect their efforts towards the development of better disease treatments.

To complete this putative scientific revolution, biomedical researchers must become skilled at extracting nuggets of knowledge from genomic databases; they must become skilled at knowledge discovery in databases (KDD). KDD was a term coined in 1989 within the artificial intelligence (AI) and machine learning research community to describe the process by which knowledge is extracted from data. "KDD refers to the overall process of discovering useful knowledge from data while data mining refers to the application of algorithms for extracting patterns from data without the additional steps of the KDD process (such as incorporating appropriate prior knowledge and proper interpretation of the results). These additional steps are essential to ensure that useful information (knowledge) is derived from the data," explain Usama Fayyad and Gregory Piatesky-Shapiro in an overview of KDD in the book *Advances in Knowledge Discovery and Data Mining*.[12] KDD is not purely cybernetic. It requires iterative interactions between a human expert (a biologist or medical researcher, for example) and a computer.

A pattern is a relationship among bits of data within a database. Fayyad and Piatesky-Shapiro, of Microsoft and GTE Laboratories respectively, define knowledge as a pattern that exceeds some user-specified threshold of interestingness, where interestingness is a function of the pattern, the validity of the data, and its novelty, usefulness, and simplicity. This definition seems very relevant to what is happening in the biomedical sciences today. For example, a *pattern* in data from a microarray experiment may be a set of 50 genes whose expression is elevated over twofold in prostate tumors relative to nontumor prostate tissue. The *validity* of the data is determined by testing the technology for reproducibility. Sampling, hybridization, and detection techniques will inevitably bring some level of variation (noise) into the results. Statistical measures should be used to track the variation introduced by the technology and can thereby help prevent inappropriate conclusions. *Novelty* is obviously important. If the same 50 tumor-induced genes were described repeatedly in the past, then this pattern would not represent new knowledge. The 50 tumor-induced genes would be *useful* if they led to a way of classifying prostate tumor types, so that specific drugs could be better tailored to each type. *Simplicity* implies the ability for human beings to understand the relationship. If the 50 tumor-induced genes remained cryptically embedded in a complex pattern involving 5000 genes, then this would compromise the ability to communicate and use this information.

One could argue that genomic analyses have not yet generated and may never generate a very significant amount of knowledge. However, I feel that the evidence provided here suggests that knowledge discovery from genomics databases will accelerate as more genomic information is transduced and as researchers become more skilled in the art and science of KDD. One could also argue that information-intensive analyses merely complement more fundamental and more important single-molecule studies. Single-molecule studies are generally still needed to confirm and expand on *in silico* findings. We also continue to desire simple, one-molecule explanations to biological phenomena, and we still seek a single compound to cure each of our ills. However, current trends suggest that advances in genomic technologies will enable more of the most fundamental messages of life to be more accurately and inexpensively transduced into a digital format. There is little doubt that information scientists will be able to continue to improve algorithms for analyzing genomic data and that biomedical researchers will become skilled KDD practitioners. This suggests to me that *in silico* work will soon become the dominant framework for biomedical studies and that the single-molecule studies will become merely complementary.

So long as it brings a deeper understanding of life and improved health care without doing harm to people, viva la Genomics Revolution!

18

The Final Act

In the spring of 2000 two teams of researchers, one from Celera and the other consisting of Human Genome Project (HGP) members, were each palpably close to sequencing the entire human genome. As the historic milestone loomed, the news media were abuzz, familiar controversies boiled up, and the endeavor took on more of the characteristics of a staged drama than of a long-distance road race. There were still plenty of zany antics, but at this time the gravity of the situation and its implications for biomedicine and for future business, legal, and health care policies and practices seemed more pertinent than before. And so familiar voices were called forth as the curtains rose for the final act.

The scene begins in Washington, D.C., where the legislative branch of the U.S. government plays host to a continual stream of distinguished guests. Here, experts, advocates, leaders, and spokespeople from all walks of life and representing all sorts of interests come to express themselves before committees of elected officials, reporters, and public cameras. It is part of an ongoing exchange that embodies the ideals of a democratic and representative government. The U.S. Congress provides a public forum for discussing many of our society's most pertinent ideas and policies, a stage for debating some of humankind's most vital interests.

In April of 2000 a proud but indignant figure took the stage; Craig Venter was once again before a subcommittee of the U.S. House of Representatives' Committee on Science to clarify and defend Celera's position on human genome sequencing. In his prepared statement Venter explained how he and his colleagues had proved the naysayers wrong. They had verified that Applied Biosystems' (their mother company's) new fully automated sequencers worked as intended; shown that these machines could

201

help advance the completion date for the HGP; demonstrated that the shotgun sequencing method could work on organisms with large and complex genomes; and confirmed that a cooperative effort between industry and government could bring highly accurate sequencing data to all researchers, quicker and at less cost than would otherwise be possible.

The 120 million base-pair fruit fly (*Drosophila melangaster*) genome had been shotgun sequenced to near-completion and in record time through a cooperative effort of academic scientists, led by Gerry Rubin of the University of California at Berkeley, and Celera scientists. The team published its findings and made the sequences available to all. Although there were still over 1000 gaps in the sequence, the fly genome researchers throughout the world reveled in this new wellspring for knowledge discovery. At the congressional hearings Venter suggested that only the recalcitrance of the HGP officials prevented such communion on the human code. He then snubbed his detractors further by characterizing their actions as despondent and glory-seeking. HGP officials had spent billions of taxpayer dollars, failed to incorporate the most efficient sequencing strategies, and now appeared to be so concerned with saving face that they were ready to let cherished "quality and scientific standards" be "sacrificed for credit."[1] Meanwhile, "since the Congress began funding the human genome effort over five million Americans have died of cancer and over a million people have died because of adverse reactions to drugs."[2] With such righteous indignation one might believe that this was neither a madcap race nor a dramatic play, but nothing less than a Holy War.

Efforts to enjoin the two largely redundant sequencing efforts had failed only a few weeks earlier. A powwow between representatives of the two groups in March 2000 had ended in acrimony and the Wellcome Trust, a large backer of the HGP, publicly released a letter detailing complaints over Celera's stance in a proposed joint effort with the HGP. The HGP letter nastily suggested that it was Celera's intent "to withhold information and delay progress." The human sequences held far more monetary value than the fruit fly sequences, and Celera's terms for sharing human sequence data were therefore considerably more stringent. Researchers would have to agree not to redistribute the data to others or use it in a commercial product without prior approval from Celera. Venter stated that he and his coworkers were "unapologetic in seeking a reasonable return for our efforts." Needless to say, it was not the same "free-access for all" vision that HGP participants passionately demanded. While negotiations were still underway with Celera, the National Institute of Health and the Wellcome Trust solicited bids on behalf of the HGP for contractors to supplement the HGP's sequencing efforts and thereby beat Celera.[3] Their favored partner appeared to be none other than Celera's arch rival, Incyte, which had maintained a low profile during this time with much more of the company's efforts focused on the valuable gene sequences and clones than on genomic sequencing.[4] The HGP maneuvers did not amount to much, though they did serve to increase the rancor of Celera officials.

Actions at the U.S. Patent Office were also fanning the flames of controversy at this time. One particular issue was a 1995 patent application filed by researchers at Human Genome Sciences, Inc. (HGS). HGS scientists had discovered an EST (partial gene sequence) with sequence similarity to a class of proteins known to be expressed on the surface of cells. They obtained the sequence of the entire gene and determined that it was expressed in cells of the immune system. BLAST and other computer algorithms suggested that the gene had strong similarity to G-protein chemokine receptors, a group of cell-surface proteins known to act in eliciting an immune system response. In April of 2000 HGS was awarded a patent on the gene sequence, which the company had named HDGNR10. The trouble was that HDGNR10 was identical to none other than CCR5, the coreceptor for the AIDS virus (discussed in Chapter 11). Researchers who had independently discovered CCR5 were outraged. They had demonstrated CCR5's function in AIDS through laboratory experiments at the U.S. National Institute of Health and published their findings in 1996, yet it appeared that their difficult and scientifically rewarding studies were now being overshadowed by "armchair" biology.[5] An HGS press release announced the issuance of the "CCR5" patent, and HGS stock promptly rose 21 percent.[6]

The USPTO had issued patents on hundreds of genes that, like HDGNR10, were known by little more than BLAST and other *in silico* analyses. However, the validity of these patents had never been tested in court. In 2000 the USPTO had a backlog of about 30,000 additional gene patent applications of this type, and even before the CCR5 controversy began the patent office was being pressured to alter its policy.[7] Many scientists still felt that industrial-scale and computer-driven studies were significantly less substantive than "wet lab" or single-gene studies. In opposition to the use of bioinformatics as support for gene patents, Francis Collins, director of the HGP, stated that "these are hypotheses. They could be right or they could be wrong."[8] Of course, animal-based, single-gene, "wet lab" studies can also yield wrong assertions, and one never really knows whether a new medical application has been invented—regardless of its patent status—until it is applied to a large number of patients. Nonetheless, the USPTO suggested a change in policy that would require applicants to show a "substantial, specific, and credible" use for a gene sequence. This does not resolve the issue, for there is little agreement on the meaning of these three words, but it does signify an attempt to change the direction of the pendulum and make it harder for patent applicants to obtain gene patent claims.

Many people did not want the historic significance of the completion of the human genome sequence to be tarnished by the dispute between the commercial and noncommercial sequencing teams. Parties from near and far sought to broker peace between the two sides, including Eric Lander of the Whitehead Institute, Norton Zinder and Richard Roberts of Celera's scientific advisory board, Ruth Kirschstein of the National Human Genome Research Institute, Ari Patrinos of the U.S. Energy Department's human genome team, and Donald Kennedy, the editor of *Science* magazine. Their efforts led to a series of meetings between the two sides over the course of several months and then finally to a truce.

On June 26, 2000, both parties took part in a hastily planned ceremony at the White House commemorating the completion of "the first survey of the entire human genome." This event was rather contrived and somewhat ironic, for there were really two

genome surveys and each was still short of its respective goal. At
that date the HGP group, which was hoping to announce a 90 percent complete "draft," had sequenced only about 85 percent of the
genome. HGP had stuck to a tiled approach, where overlapping
fragments (BACs) were sequenced one by one (the BAC by BAC
approach). Celera had sequenced only one person's DNA an average of 4.6 times over, using shotgun sequencing. The company had
aimed to sequence five or more individuals with tenfold coverage.
At that time each survey was riddled with gaps. Celera's biggest
contiguous sequences were less than two million base pairs in
length. Clearly, there was more sequencing to be done. Nonetheless, nearly everyone wanted to take advantage of the brief thaw in
relations between the two sides to exhaust the theatrics and refo-
cus on what should be the primary interest, the analysis of the preliminary data. There would not be a large-scale collaboration
between the two sides. Instead, the two groups simply agreed to simultaneously publish independent reports in the fall of 2000.
There would thus be no declared winner.

Celera scientists estimated that the genome consists of 3.12
billion base pairs of DNA, while HGP scientists pegged it at 3.15
billion base pairs. The Celera version of the human genome is initially available to only a select set of researchers, while the HGP's
human genome is accessible by everyone. The task of identifying
all the genes, interrelating the genes, and determining their function and their relationship to the traits and diseases of interest has
really just begun.

One may wonder how this drama would have played out if genomic companies had never come into being, if all DNA sequence
patents were banned, or if only an elite set of scientists effectively
controlled research funding and the publication of research findings. It might have taken longer, but the human genome would
have still been sequenced in its entirety and the genes would have
been found. If the pursuit of the human genome sequence were a
purely academic endeavor there may have been less of the crass
hype and one-upmanship endemic to the world of business and
typified by Venter's speech before the congressional subcommittee. The public would probably have seen a much more harmo-

nious scientific community, which would in turn have fostered greater faith in our scientific institutions and leadership. Our society would then suffer fewer of the ill effects, the conflicts, and controversies that inevitably emerge as technology and science race forward. However, in considering how we wish to shape the future of health care and biomedicine we should fully consider all benefits and costs of our actions. Societal dissonance also underlies the impetus for creating the new scientific paradigm and the new industry described here. Many people believe that our current state of biomedical knowledge is inadequate, and for many people genomics presents great new hopes for the future.

19

Future Prospects

Catalyzed by public and private genome sequencing initiatives, propelled by an ongoing desire to overcome sickness and disease, and fortified by the economy, technology, and optimism of the Information Age, we now enter a new age in biology and medicine. We hold great hopes, yet we contend with a great deal of uncertainty and not a little trepidation over what this new age may bring. "Information technology enthusiasts like to say that the digital revolution is as profound a change as the industrial revolution. But the genomics revolution, the turning of information technology on ourselves, may be more transformative still," one magazine editor has noted.[1] We are all aware of the power of information technology and the relentless change that it brings to human society. It makes many wonder where genomics, this new, inwardly-directed information technology, will bring us.

We can anticipate that the trickle of new innovations that are starting to have an effect on medicine and medical research will increase. We should see more diagnostic tests for single-gene genetic diseases; more gene variants linked to complex genetic diseases; more clinically valuable gene products and new targets for drug development; greater use of gene expression microarray analysis in drug discovery research and disease diagnosis; and further retooling of the drug development process for greater efficiencies and higher throughputs.

Disease definitions will be refined by powerful new diagnostic tools that assay key mRNAs, proteins, and/or other molecules. Consequentially, there will be many new disease categories. For example, there may be hundreds of distinct types of asthma, arthritis, or dementia, each recognizable through the application of new, simple, and low-cost tests. Patient populations will be segregated further by genotype, specifically by the gene variants that affect disease progression and treatment response. New treatments, existing treatment options, and even previously discarded treatment choices will be tailored to these newly defined patient populations. The result will be more effective and more individualized treatments.

The genomics revolution does little to rid society of the disparity between those with money and power and those without. New treatments may be costly at first and limited to wealthy individuals and those supported by well-funded health plans. However, costs will fall as patents expire and as alternative treatments emerge. So long as monopolies and cartels can be avoided, the same economic forces that have reduced the price of computer chips and electronic gadgets will bring cost savings to health care consumers. Already, for example, the prices of gene expression microarrays have fallen considerably since they were first introduced for biomedical research. DNA chips that once cost $4,000 could be obtained for $400 only four years later. Economies of scale, engineering and manufacturing improvements, and competitive pressure will likely result in further discounts. It is not hard to imagine that one day diagnostic DNA chips might be as inexpensive and as routinely applied as common blood and urine tests. Genomics is hastening the development of new medicines, reducing the overall cost of treatments, and will eventually bring new disease prevention and treatment options to nearly the entire human population.

Despite the growing arsenal of patents and the multiplying battalions of commercial interests, there is little doubt that rapidly increasing quantities of genomic information will be free and widely available through the Internet. Patents and corporate greed are not stopping this trend nor are they hindering research. After U.S. President Bill Clinton and U.K. Prime Minister Tony Blair issued a joint statement in March of 2000 backing free access to human DNA sequence data, the response from genomics companies was swift and telling.[2] Did the for-profit genomic concerns register their protest to the world leaders, greedily defend the broadest interpretation of their intellectual property, or pull out all stops to defend their stock prices, which were rapidly plunging in response to the statement? Absolutely not! Instead, every major genomics company issued a press release praising the Clinton/ Blair statement! "The President is dead right in opposing broad patents of the human genetic code," declared William Haseltine of Human Genome Sciences.[3] "Ensuring that this information is equally accessible to all is good for our business and that of the entire biotechnology industry," stated Roy Whitfield of Incyte. Indeed, not even Microsoft chairman Bill Gates (were he the head of a genomics company) would have dared defy the government leaders' seemingly offhand and innocuous statement, because when it comes to fundamental biomedical information the public will is unyielding. Genomic data will be available for further research and science will not be held hostage! Commercial outfits may fight over material goods such as the drugs and diagnostic tests derived from genomic data, but the information itself and health-related knowledge derived from it does not stay contained. (Freshly discovered genomic data might be patented and sold, yet like so much other information, in short order it ends up on the Internet free for all to use, except for resale.) The war against ignorance and disease supersedes lesser squabbles over profits and fame. Already, many of the commercial firms, such as DoubleTwist, Incyte, Lion Biosciences, and HySeq, are placing their raw data and bioinformatics software on the Internet to demonstrate their good will and to entice researchers to learn of and purchase additional services.

It is also likely that your individual genomic data will one day be accessible to you and your designated physician. The tools for accessing this information are rapidly advancing, and genomic companies such as Celera are promising to make this information available to individuals over the Internet.[4] We will then be able to analyze our own genetic makeup and evaluate our personal disease risks associated with various lifestyle or treatment choices. (This will, no doubt, put great pressure on the health insurance companies, which presumably would not have access to this information and thus could not evaluate their financial risks in their accustomed manner.) Internet links may guide individuals to additional information on relevant diseases and traits, physicians, patient groups, and so forth. The idea is to extend the breadth of Internet-driven personal empowerment to include the ability to design highly sophisticated and utterly personal health care solutions.

Increased knowledge of our internal information network will also enhance our ability to bypass pharmaceutical-based routes to health care and rely instead on personalized and improved disease prevention measures and/or alternative treatments. The cells of our bodies respond immediately to environmental inputs, including material introduced through eating, breathing, and drinking, as well as other sensory inputs. Genomic technologies will allow us to hear these messages and take corrective actions well before disease becomes manifest. They may come to be like new senses, finely tuned to numerous currently undetected health indicators.

Knowledge of our internal information network will come mostly from an explosion of new genomic database analyses. A growing army of mathematicians and information scientists will develop increasingly powerful and more useful algorithms and computational processes for finding biomedical knowledge in these databases. A growing regiment of biologists and medical professionals with training in mathematics and information sciences will lead these knowledge discovery missions. New recruits are now being lured from other pursuits with six-figure incomes and the promise of a chance to make a real difference, to bring significant and positive change to society.

The amount of data that knowledge discovery practitioners have to work with will continue to increase, and the quality of this data will continue to improve. More of the most fundamental messages of life are now being wholly intercepted and transduced through new genomic initiatives. For example, in February of 2000 a collaboration between researchers at CuraGen Inc. and the University of Washington resulted in a comprehensive analysis of nearly all protein-to-protein interactions in yeast.[5] To determine which proteins are communicating with each other, every yeast protein was tested for its propensity to bind to other proteins. Plans are underway to assay human proteins in this way and, of course, to store the information in an electronic database with links to other protein data. Equally impressive plans are in the works elsewhere. They include intensive efforts to uncover more information, such as regulatory and structural signals, hidden within DNA and amino acid sequences, and efforts to amass databases of comprehensive gene transcript and protein expression information from a wide range of different tissues and cell cultures. The maturation and linking of these databases will facilitate extraordinary new medical advances.

A massive database network can be analyzed in countless ways. The possibilities are mind-boggling. Imagine a matrix with column headings listing every imaginable tissue sample, from every imaginable source, treated with every imaginable compound. Imagine rows listing every known gene transcript, every protein, and every imaginable measurable characteristic of the sample: the age of the donor; the status of every variation in his, her, or its genes; the environmental conditions to which he, she, or it was exposed; and characteristics of his, her, or its health. In the matrix itself would be numerical values corresponding to these measurable characteristics. Sound far-fetched? Consider the resources being amassed by the Icelandic genomic company DeCode Genetics. The people and government of the close-knit nation of Iceland have bestowed upon DeCode, a 1996 start-up, exclusive rights to nationwide databases of extremely detailed and comprehensive health and genealogy information (encrypted to help maintain privacy rights). The company also has partnerships with Icelandic hospitals, the pharmaceutical giant Roche, and dozens of smaller firms. Similar alliances are being forged elsewhere.

Now suppose that this colossal set of genomic information was readily available to all researchers. Knowledge may rain down from this hypothetical mother of all biomedical databases. Everything from the link between cigarette smoking and lung cancer to the link between globin variations and sickle-cell anemia to countless new molecular associations and pathways could be found within such a database, including complex associations involving multiple factors. These associations could be replete with statistical measures of certainty, since the requisite controls could also be incorporated in such a database. And if this database includes detailed and accurate records of molecular communications, such as a complete set of protein-to-protein interactions or the molecular response to a series of thousands of independent gene variations, then the findings could be as hard and true as any lab-based single-gene study. The foundation would be set for an era of knowledge discovery that would exceed that of Alexandria two thousand years ago.

This foundation is being laid now. The human genome sequence is just one of the first blocks. Many more pieces will follow and together they will support a tremendous new knowledge base and great health care advances. Who is the architect for this structure? Who will lead the development of the mother of all biomedical databases? Will it be directed by an elite panel of experts, a business titan, or the new president of the United States? Will Incyte, Celera, DeCode, DoubleTwist, or some other company establish the world's most authoritative source of genomic information and attain monopolistic control over health care data? Will the U.S. Congress remove incentives for privately controlled databases and grant full authority over genomic data to a nonprofit institution? The answer to each of these questions is "No." No one authority will build the foundation for future health care advances (see Fig. 19.1). No one will create the mother of all databases. It will emerge, instead, from disparate sources lacking in preestablished order or organization. It will be a network of linked databases, imperfectly compatible and thus somewhat less than any one authority's idealized vision. The genomic database Web, the wellspring of new biomedical knowledge, will always be the sum of many independent ambitions and the marvelous synergy of those efforts that happen to succeed.

Financial support for genomic initiatives has risen dramatically. Approximately $80 billion in private investment was raised in the first three months of 2000 alone.[6] Thousands of new genomic companies have been launched, each with ambitious goals for either creating new tools for accessing genomic data, building a new genomic database, linking databases in better ways, creating better knowledge-discovery tools, or deriving new medical diag-

Institution	# of issued U.S. patents as of spring 2000
U.S. Government	388
Incyte Genomics	356
University of California	265
SmithKline Beecham	197
Genentech	175
American Home Products	117
Isis Pharmaceuticals	108
Massachusetts General Hospital	104
Human Genome Sciences	104
University of Texas	103
Institut Pasteur	101

FIGURE 19.1   No single institution owns the rights to a majority of known human genes. *(Data from PriceWaterhouse.)*

nostics or therapeutics from transduced data. These commercial ventures have not put off publicly funded projects. On the contrary, government and foundation support for genomic initiatives have risen dramatically.

Academic institutions are providing further support for the new information-based approach to biomedicine. In 1998 Harvard University earmarked $200 million for a new Genomics Center that will support about a dozen faculty research leaders drawn from a variety of scientific disciplines. Similar research units have been formed at Princeton University, the Massachusetts Institute of Technology, the California Institute of Technology, and many other leading academic institutions. In 1999 Stanford University accepted a $150-million donation from Jim Clark, one of Silicon Valley's most successful entrepreneurs, to help create its new Center for Biomedical Engineering and Sciences. Stanford administrators proclaimed that biology is now a "foundation science" and a sufficiently powerful magnet for attracting significant cross-disciplinary scientific and engineering work.[7] "Biology is not just for biologists," U.S. National Institute of Health (NIH) director Harold Varmus declared in 1998 as the NIH planned new genomic technologies and related "bioengineering" initiatives that paralleled those of Stanford.[8]

The University of Washington in Seattle and its supporters have undertaken a series of initiatives, each bolder than the last, and each designed to place the U.W. and Seattle at the forefront of the genomics revolution. Leroy Hood, an extraordinary inventor and a magnet for talent, was recruited to be the William Gates III Professor and Chairman of the Department of Molecular Biotechnology. Maynard Olson, the medical geneticist who I heard lecture at Stanford on large-scale sequencing in the late 1980s, Philip Green, a computer expert who had developed key algorithms for interpreting genomic sequence data, Mary Claire King, a medical geneticist who had discovered several important disease-causing genes, and many other talented scientists have joined the new department. A Human Genome Project sequencing center was established. Then, in the spring of 1999, Hood announced the creation of a new cross-disciplinary Institute for Quantitative Systems Biology. Here, biology was meant to be transformed from a descriptive science to a quantitative, predictive science, and not only would engineers, mathematicians, and computer scientists be called forth, but the engines of industry would also be summoned to the cause.[9] Collaborations with small companies were to be encouraged so as to allow the Institute to have access to new technologies. In exchange, the Institute would make its innovations commercially available to these corporate partners. Remarkably, within a year Hood decided to quit the U.W. and spin off the planned Institute, making it independent of the university, which despite the great strides that it took to be at the cutting edge, was seen by Hood as too bureaucratic and slow-moving.[10] Meanwhile, the U.W. did not hesitate in creating yet another genomics-based group, the Medical Center's Cell System's Initiative (CSI), which is "dedicated to the comprehensive study of the information systems that operate within all living cells and organisms."[11] The biotech firm Immunex signed up as a corporate sponsor for CSI, promising to provide $2.1 million over three years, and collaborations have been established with leading computer scientists and electrical engineers. No other community seems to have made

such ardent strides in forging an incubator for genomic research
and industry as Seattle. However, their efforts are clearly indicative of both the widespread hopes and desires for the Genomics
Age and the determination to do everything humanly possible to
realize these dreams.

Naturally, each of the new academic programs includes an educational plan. New crops of scientists are to emerge from these
programs with much stronger computation and engineering skills
and a more profound appreciation of the information-intensive
nature of living organisms than have prior biomedicine graduates.
They will feed the growing genomics industry and seed the growing academic rosters of universities throughout the world.

Unlike most of their industry counterparts, students and other
university researchers are free to pursue genomic research that
is not directly linked to the near-term development of medical
products. In what ways will the new legions of cross-disciplinary researchers apply the new and ever more extensive genomic databases? Perhaps, they will uncover the "logical structure" that
underlies organismal development as Norman Anderson, Leigh Anderson, and other pioneers in genomics had imagined decades ago.
The Andersons and other logical positivists felt that complex
process of development can be solved via an approach "based on
the methods of artificial intelligence and expert systems" applied to
a sufficiently large genomic database.[12] If we knew all the molecules of a cell type or organism at each stage in its development,
then, presumably, we could reverse-engineer the system. The engineers at Compaq Computer Corporation completed a somewhat
analogous task in the 1980s when they figured out how to make a
computer that was compatible with software and hardware developed for the IBM personal computer. Compaq employees could
not simply copy the circuitry of the IBM computer, for that would
infringe IBM patents. Instead, they meticulously tested the information processing system, measuring its outputs in response to a
wide variety of inputs, and used this data to deduce the underlying
logic of the IBM machine. This knowledge was then used to design
a replica, the first of many enormously successful IBM computer
clones. Reverse-engineering the circuitry that regulates gene ex-

pression would provide a predictive model for how perturbations in the system affect cell responses and would represent a far more comprehensive understanding of the blueprint for life than the cracking of the triplet code that dictates protein translation.

There are, of course, medical benefits to having a predictive model for the network of molecular messages that forms the basis of cell function and dysfunction, and so corporate research groups will also be positioned to make contributions to efforts to reverse-engineer living systems. Such a comprehensive model would help in pinpointing perturbations that lead to either disease or to health, and one could then derive logical, information-based remedies. These messages could lead one to particular changes in one's diet or lifestyle. Or the digitized remedy could be translated into a molecular concoction, a new prescription drug or a fittingly complex combination of drugs. So long as life is molecular, drugs will be molecular, and just as an electronically derived chess move can be converted into the movement of a real chess piece on a real chessboard (by a person who reads the computer's instructions), so too can a digital drug be converted into real drug candidates. Increasingly rational approaches to drug design are in the works. The Menlo Park start-up Entelos, for example, is developing sophisticated electronic models of disease pathways that will help pinpoint the best molecular targets. Other firms specialize in designing new chemical compounds to fit electronically created models of these target molecules.

There is little doubt that increasingly sophisticated electronic models of living systems will aid in understanding their logic. These efforts may also aid in understanding our mental and technological limits. Systems with millions of interacting parts are incredibly complex and may be beyond the scope of any imaginable computer system. Nonetheless, we can expect that bright minds will continue to try to model living systems and that something at least close to an electronic clone of a living organism or a *de novo* electronic creation will prod many of us to further contemplate the question "What is life?"

Genomics researchers may also uncover new insights into the features that distinguish humans from all other species. According to most anthropologists and linguists, human language skills are among our species' most distinctive features and are closely associated with our seemingly unmatched cognitive skills.[13] Although culture determines which particular language we speak, the underlying communication skills are both unambiguously innate and unambiguously unique to humankind. Children deprived of external language will invent one on their own, and, whereas other creatures communicate and may apply rudimentary rules for determining the form and meaning of symbols, none have a grammar that comes close to matching the enabling capabilities of those used by people.

The Holy Grail for many linguists is the delineation of the core generative grammar, the communication constructs that are universal and exclusive to humans, the so-called Universal Grammar. Presumably, this environment-independent grammar is encoded in our genes, which somehow oversee the development of certain critical, though still poorly defined, neuronal circuitry. Cracking these codes—the genes to neuronal circuitry code and the neuronal circuitry to Universal Grammar code—will therefore reveal the secrets of one of humankind's deepest mysteries.

Already, the existence of a gene that has an integral role in human grammar has been inferred through an analysis of families with a high incidence of a specific disorder in language development that renders its victims unable to learn simple grammatical rules, though they are otherwise normal. Meanwhile, dozens of genes that act informing neuronal connections have been discovered through animal studies, and the neuronal circuitry responsible for language in both nonhuman and human species is slowly being uncovered. In the not-too-distant future genomic studies are likely to uncover much more of the complex network of molecular messages that give rise to neuronal structures, including those that enable human language.

Finally, by deciphering the molecular messages of life, genomicists may not only shed light on the question of *how* humans are different from other species, but they may also discover *why* we are different, a topic that elicits vigorous debate among linguists, anthropologists, biologists, and many others. Many evolutionists presume that the human language faculty and other cognitive functions are adaptations specifically acquired or designed through the process of natural selection. If this is the case, then telltale DNA sequences should exist for each mental function. We should be able to map structures that specifically enable human language, for example, to particular DNA sequences and trace their origins, as has been done for many of the genes that enable hearing and sight, for instance.

However, many people—including Noam Chomsky, the linguist most responsible for the current conception of an innate Universal Grammar—appear uneasy with such a Darwinian explanation for human language.[14] Chomsky has contended that humankind's unique Universal Grammar may be the consequence of human brain structure, but not the driving force behind its evolution. This notion is echoed by many complexity theorists. Stuart Kauffman and others have drawn attention to the self-organizing and emergent properties of complex systems. The Universal Grammar and other human cognitive functions may be the natural consequence of a more generally selected advanced neuronal network. Again, telltale signs should emerge through genomic research. In this case, DNA sequence evidence may show support for greater neuronal plasticity or other general anatomical features. Furthermore, and perhaps more significantly, genomic research should better model the complexity of the internal human information network and show evidence of some of the features of these emergent properties, which are more the natural product of mathematics than of Darwinian selection.

Alternatively, in modeling complex biological systems genomicists may hit the proverbial brick wall, some self-discovered set of limitations on how much we can know and predict about ourselves. Mathematicians hit such a wall in 1931 when Kurt Gödel proved that axiomatic formulations of number theory always included undecidable propositions, his famous Incompleteness Theorem. According to Gödel's theorem, no mathematical framework could both be entirely contradiction-free and provide a foundation for deriving all the complex rules that whole numbers (1, 2, 3 . . .) obey. The study of complex systems, such as those involving millions of messages and information processing units, may face similar obstacles. Proof of these limitations, should they exist, will no doubt delight those that feel that genomic research aspires to feats that are too lofty for human beings. Nonetheless, biomedical researchers have yet to discover such limits, and, as the events described in the preceding chapters indicate, at this time the collective will to carry on this work appears quite strong.

Last, biomedicine's new information-intensive approach begs for a fresh look at questions of how life began and where it is going. Radiation and other energy sources provide a means for randomly modifying the internal messages of life and natural selection and other factors provide a means for their evolution, but where did the first message come from and what did it say? Is there an ultimate source for the messages of life?

One inspiration for the story printed here was a book written by Professor Werner Gitt of the German Federal Institute of Physics and Technology in Braunschweig, Germany. In *In the Beginning Was Information*, Professor Gitt first nicely explicates biology's informational roots and then contends that God is the source of all messages and the "natural laws about information fit completely in the biblical message of the creation of life."[15] Another perspective is provided by Professor Werner Loewenstein, presently director of the Laboratory of Cell Communication at the Marine Biological Laboratory in Woods Hole, Massachusetts, in a recent book on the flow of molecular information. In *The Touchstone of Life*, Loewenstein contends that the simple and unique properties of a steady stream of photons provide the nourishing message for life's genesis.[16] Indeed, our artistic and spiritual selves continue to listen for and seek to decode messages that emanate from nonliving sources. I am reminded of these lines from a song:

> Cloudless everyday you fall upon my waking eyes
>> Inviting and inciting me to rise
>> And through the window in the wall
>> Comes streaming in on sunlight wings
>> A million bright ambassadors of morning

And no one sings me lullabies
    And no one makes me close my eyes
    So I throw the windows wide
    And call to you across the sky.[17]

I am not sure about the ultimate source code, but I am certain about two irrepressible messages. One tells us to live long and healthy lives and the other beseeches us to know and express ourselves.

Appendix

HOW THE PATENT PROCESS WORKS

What is a patent? It is the right to exclude others from producing or using a discovery or invention for 20 years. It therefore gives a temporary monopoly to the owner of the patent, which typically is the institution for which the inventor(s) work. Government patent offices award patents, and patents are enforced by courts of law.

What is the purpose of patents? The patent system is designed to promote the development of innovative new products and services. It does this by providing an incentive, the 20-year monopoly. Companies and investors are more willing to support research and development when there is assurance that the fruits of their efforts will not be undermined by imitators. Competitors may direct their resources towards inventing a better product or process.

Presumably, if they are not self-employed, inventors are given an incentive by the institution for which they work. Some universities, for example, grant a substantial portion of license revenues to inventors. Companies vary in their policies. With small companies, inventors are often among the company founders and own a significant share of the company.

Are there other ways to promote innovation? Absolutely! Universities have been wellsprings of innovation, yet until recently patent rights have not been a significant motivating factor. Academic scholarship has its own rewards. However, universities are not in the business of transforming innovations into useful products, and government institutions have not excelled at it. In terms

221

of bringing new products to the marketplace, nothing has come close to matching the efficiency of profit-seeking companies.

What can be patented? An invention may be a "composition of matter" (material, such as a safety razor, an AIDS-abating drug, or a flat-screen computer display) or a method (such as the chemical steps used to extract an analgesic drug from a medicinal plant, the method for filtering out static noise from magnetic tapes, or a particular technique for killing bacteria in fresh fruit juice). In order to receive patent protection an invention must be novel, nonobvious, and useful. "Novel" means the invention must not have been previously described. By U.S. law the date that the invention is first fully documented (as in a lab notebook or in the patent application itself) is the date used for determining who was first to invent. Furthermore, an invention must not have been described publicly (such as in a public database, in a scientific journal, or at a conference) more than a year prior to the patent application. Outside the United States, most countries deem the date that the application was "filed" (submitted) as critical in determining who has priority in obtaining a patent. "Nonobvious" means that at the time of the invention people familiar with the field would not have thought of the invention. The "useful" requirement is meant to exclude inventions without an intended use or with only inane uses. There are a few additional requirements: The applicant must adequately describe how to make the invention, present the best known mode of the invention, and pay certain fees to the patent office. Some things have become specifically excluded from patenting—including human beings, naturally occurring articles (as they occur in nature), processes that can be performed mentally, and medical procedures, such as a technique for heart bypass surgery. Natural processes cannot be patented.[1]

Can an animal be patented? A plant? A microbe? Cells or other material derived from a person? Currently, certain nonhuman organisms can be patented. In 1980 the U.S. Supreme Court ruled that bacteria that had been genetically altered so that they were able to help clean up oil spills could be patented.[2] In 1988 Harvard University was awarded a patent on a genetically engineered strain of mice.[3] Since then, quite a few animal strains have been developed that mimic human diseases and are particularly useful for evaluating possible new disease treatments. Many organisms have been patented. Thus, a precedent has been established in the United States for allowing patents on novel, useful, and nonobvious genetically engineered organisms. However, such a precedent has not been established in most other countries.

Can genes be patented? Yes. In 1991 a federal court of appeals upheld broad protection to the erythropoeitin gene[4] and in 1993 a court decision assured broad rights to the beta-interferon gene.[5] Genes are classified as chemical entities by the U.S. Patent and Trade Office (USPTO) and can be claimed as compositions of matter, so long as their DNA or protein formulations are novel, useful, and so forth. Methods for using the gene may also be claimed. Typically, claims in gene patents cover the purified gene and its encoded protein, cells or organisms that are engineered to produce the gene or its protein, methods to purify the gene or protein, and the use of the gene or protein to detect or treat particular diseases or conditions. Genes or proteins as they exist in nature, in the cells and tissues of the body, are beyond the scope of these claims.

A technique developed by a small biotechnology company in Cambridge, Massachusetts, is the center of an interesting new gene patent debate. Transkaryotic Therapies Inc. (established in 1988) has developed a proprietary method for altering the expression of genes in their endogenous state (as they naturally exist in cells). One of their lead products-in-development is the erythropoietin protein, which is harvested from gene-activated cells. Erythropoietin, a red blood cell booster, also happens to be Amgen's lead product, yielding the company greater than $1 billion in sales in 1998. Amgen derives its erythropoietin from genetically engineered cells grown in large bioreactors and has patents with broad claims on the gene and its encoded protein. By utilizing the endogenous gene, rather than a recombinant gene, Transkaryotic Therapies may be circumventing Amgen's patents. With billions of dollars at stake, it is no surprise that Amgen is suing Transkaryotic Therapies for patent infringement. The outcome of this litigation may have important implications for a number of companies that are developing drugs that target the regulatory regions of endogenous genes. Gene regulatory sequences, which are being identified and patented at a much slower pace than protein encoding sequences, may turn out to be very valuable.

How does one get a patent? What is the process? One can readily imagine that the patent-worthiness of an invention may be subject to considerable debate. Similarly, the degree to which an invention should exclude others, known as the *scope of the claims,* may also be hotly contested.

The process of arguing with examiners of the patent office over whether an invention or discovery is worthy of a patent and the degree to which it should exclude others is referred to as patent *prosecution.* The patent has two parts. One is the *specification,* which describes the invention or discovery, how to make or obtain it, and its novelty, nonobviousness, and usefulness. The other part, the *claims,* consists of carefully worded sentences that spell out the items or actions that are being protected. After the patent *issues,* that is after it has been *allowed* by the patent office and fees have been paid, others whose actions overlap or *read on* those described in the claims will be *infringing* on the patent and are thus susceptible to a lawsuit.

Patent prosecution is a debate between the inventor, or more often an attorney or patent agent representing the inventor, and a patent examiner. Typically, an applicant will seek an array of claims that confer broad protection. For example, suppose that you have developed a flat-panel television that is thinner than anyone else's. You may claim (1) the technology used to make the very thin flat-panel television, (2) any very thin flat-panel television made by this technology, and (3) all very thin flat-panel televisions, regardless of how they were made. The examiner typically rejects some or all of the claims in a first *office action.* In subsequent communications the applicant may seek to change the examiner's mind and may amend the claims or write new ones. Eventually, a final decision is made. You may walk away with a patent that includes the first two claims to the flat-panel television patent application, but not the third. The process, from filing to issuance, most often takes between one and six years.

Ah, but one need not end here. Whether it is allowed or not, the original patent application can act as a starting point for additional patent applications. New claims can be amended to the same specification in patent applications known as *divisionals* or *continuations*, provided there is some support for these claims within the specification. If it turns out that your technology is also suitable for making flat-panel computer monitors, then you may able to receive rights to such uses through a divisional or continuation patent application. In relying on the specification of the original, "parent" patent application, public disclosures that would otherwise be disqualifying may be sidestepped, and an early date of invention may be retained. An old specification may also be combined with new material and used to support a new set of claims in an application known as a *continuation-in-part*. These types of applications may issue a decade or more after the parent application was filed. However, a 1995 ruling prevents the 20-year life of a U.S. patent from being repeatedly reset by these "new" patents.

In the United States the patent application and prosecution remains private until after the patent issues, while in most other countries patent applications become public 18 months after filing. U.S. policy may soon change to conform with these other countries.

How does the patent office decide who was first to invent? Patent applicants must provide examiners with information on anything known to resemble the invention or its components. The examiner also searches for such *prior art* and considers it when evaluating the novelty and nonobviousness of the invention. If two or more applicants are seeking patents on similar inventions, even after one patent has issued, an "interference" may be called. This activates a detailed review process. In Europe the issuance of a patent may give rise to an *opposition*, an additional review process in which interested parties may dispute the integrity of the patent applicants' claims. In both circumstances the respective patent office ultimately makes the call, although it may take several additional years.

What about cost? The cost of obtaining and maintaining a patent can be considerable. Most molecular biologists rely on dedicated patent agents or lawyers to write and prosecute their patent applications. Their fees and the fees that the USPTO demands adds up to a hefty sum.[6] (In recent years the total USPTO fees collected have been tens of millions of dollars greater than their operational costs. The excess has been absorbed by the government. Thus, the USPTO, which issues over 160,000 patents a year, has been a revenue generator.) Ten thousand dollars seems to be an often-quoted estimate on the total cost of getting an invention patented, assuming that the invention is patent-worthy, there are no interferences, and a patent agent prosecutes the application. There are significant additional costs for maintaining a patent once it issues and for filing applications in additional countries. Litigation costs, for either the suer or the defendant, typically run into the millions or tens of millions of dollars. Damage awards may be considerably more. The temporary monopolies that patents win have generated profits that range from $0 to over $10 billion, with the majority clustering near $0. Patents, whether *pending* (in prosecution) or issued, can be used as leverage to raise capital for starting or expanding a business, establish a market position, or dissuading competitors (see Fig. A.1). Clearly, there are important business decisions to be made throughout the patent process.

Does the patent office have the final word on what is patented? Not necessarily. Although examiners can effectively kill patent claims in a *final office action*, their authority over the claims that they allow is more limited. It is up to others to either adhere to the intended restrictions or to defy them. It is up to the patent holder to either accept the actions of those in defiance or to counteract them. In lucrative, innovation-driven fields, where disputes are common, profit-thirsty competitors may have a different interpretation of the validity or scope of an issued patent than that of the examiner or the patent owner. Although in some cases a friendly handshake may settle the issue, other disagreements are either endured with stony silence or are bitterly fought over in court. If a settlement is not reached, then a judge or jury may end up either *upholding* or *invalidating* any or all of the patent claims. The written history of the patent prosecution, the *file wrapper*, is usually scrutinized in such lawsuits. It provides guidance for interpreting the claims, as well as the logic that led the examiner to allow the patent. However, loads of additional evidence may also be brought

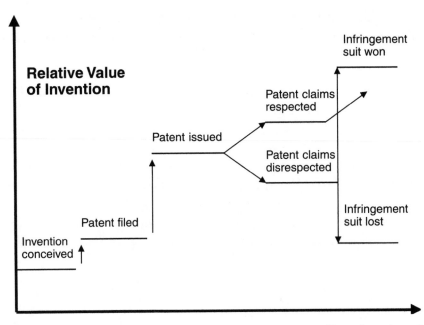

FIGURE A.1   This chart shows how patent status may affect the value of an invention.

forth. Patent lawsuits may take several years or even a decade or more to resolve.

Does patenting a gene prevent others from conducting or publishing research on it? It can, but most often it does not. For example, a search of Medline reveals that in the years since Amgen Inc. first received patent rights to the erythropoeitin gene, thousands of scientific papers have been published on the molecule. Some of the companies and academic labs that are conducting research involving the gene may have received permission from Amgen, but many probably did not. In general, most gene patent owners have not asserted rights over the use of their genes for noncommercial purposes. Once a patent position is assured, genes flow freely through the research community. (An electronic copy of the sequence is enough to commence research.) Thus, leptin, insulin, growth hormone, and dozens of other genes have brought scientific research bounty both before and after their patents issued.

Research may quickly bring on valuable discoveries regardless of the scientist's intentions. When this occurs, the patent owners may be quick to assert their rights. Therefore, for most commercial outfits a gene patent held by an external entity removes the incentive to pursue research using the patented gene. Patents usually do not remove the incentive for academic researchers to pursue research. In fact, the temporary gene lord may reward a researcher who produces an important "follow-on" discovery with a lucrative job offer, or, particularly if the new finding is likely to have additional patent potential, offer a licensing deal to the research institute where he or she works. (With regard to the previously mentioned Transkaryotic Therapies, the company did not feel that it needed a license from Amgen to manufacture erythropoietin. Amgen sued after clinical trials had begun, when it was clear that Transkaryotic Therapies intended to commercialize their version of the protein without a license.)

Patented gene sequences become publicly revealed in one or more of the following ways: After a patent application has been filed, the gene discoverer discloses the sequence in a scientific publication or presentation (and to GenBank); an independent researcher discovers and presents the sequence before a patent is issued; the USPTO reveals the sequence upon issuance of the patent; or the sequence is automatically outed through international patent filing procedures 18 months after the original application was filed. Had the USPTO not accepted gene patents, then it is doubtful that companies would publicly release their sequences. (In fact, as is discussed in Chapter 8, one company may have generated and released partial gene sequences in order to prevent others from patenting them. It is indeed a mad, mad world.)

Does a patent ensure freedom to operate for the patent holder? Not necessarily. Many patented inventions require the use of one or more other patented inventions. For example, the chemical compound in the drug Viagra was patented and claims were granted for the chemical compound itself and for the use of this compound in treating certain ailments of the heart. Later scientists discovered that the compound was useful in treating male impotence. A new patent application was filed on this new use of the compound. However, one could not use the compound to treat male impotence unless one had the right to make the compound. In this case there was no conflict, because the company that applied for the method of treating male impotence already owned the right to make the compound. However, this is not always the case. For example, PCR is a patented technique for amplifying fragments of DNA, yet many nonpatent holders have devised novel uses for the PCR technique and have received patents for their inventions. To use or sell their patented techniques they or their customers must receive a license from the owner of the original PCR patent.

Can a gene discoverer lay claim to unanticipated uses of the gene? Can a gene discoverer lay claim to uses of the gene that are not supported by traditional (single-gene) laboratory work? Herein lies the most pertinent point of contention in a heated debate over full-length gene patents. At issue is the alleged utility of newly discovered gene sequences and the scope of gene patent claims. Few scientists or others directly involved in biomedical research object to gene patent claims that are supported by experimental results that directly implicate the claimed gene in human disease. Not many object when the studies are done using a human disease model, even in something as distant from humans as yeast. However, many have strongly objected to gene patent claims that rely directly on utility arguments based on insights gathered via BLAST or other computer-run algorithms. The link between the novel gene and the purported utility is said to be too theoretical and too tenuous. Nonetheless, patent offices have granted such claims, and as a result researchers everywhere have raced to file them. (A plan to change USPTO policy and an important test case are discussed in Chapter 18.)

Could patents hinder scientific progress? Perhaps! Patents are intended to give the patent holder temporary control over the commercial use of their invention or discovery. However, in some places they may also give control over noncommercial uses as well. Although patent laws in most European countries and Japan exempt "experimental" use of patented inventions, in the United States they do not. The freedom to conduct research using patented inventions, even in universities or other nonprofit settings, is therefore unclear in the eyes of U.S. law. So far the only uses of nonlicensed patented inventions that have been judged to be acceptable are those either for "amusement, to satisfy idle curiosity, or for strictly philosophical inquiry," precisely the applications that will not lead one closer to a medical advance (at least not intentionally).[7] On the heels of the initial EST patent applications, Thomas Kiley, a patent lawyer for Genentech, warned that "patents are being sought daily on insubstantial advances far removed from the marketplace. These patents cluster around the earliest imaginable observations on the long road toward practical benefit, while seeking to control what lies at the end of it."[8] Kiley urged the U.S. Congress to make into law an exemption for experimental uses, but this had no immediate effect.[9]

One vexing issue is that there is now less of a distinction than ever between noncommercial intentions (sometimes called *basic research*) and commercial ones (*applied research*). Perhaps, it is due to the spread of the contagion known as capitalism. Potential health or agricultural benefits are used by academic and government scientists to help justify virtually every request for research funds, and there are now avenues for institutions to directly profit from research results. The new upstream-to-downstream, research-to-medicine continuum may also be due to the success of the molecular reductionist paradigm. A pathway of logical steps can often be delineated, not only between biological phenomena, but also from fundamental discoveries and innovations to medical advances (see Fig. A.2). No contributor in the pathway can afford to not be compensated. If your soccer squad won the World Cup, it would be very satisfying. However, if only the scorer of the final winning goal and the player who made the final assist received all the recognition and rewards, then you and your teammates might feel slighted or worse.

There has indeed been a tremendous proliferation of patents in the life sciences, and, even though only a few EST patents have been issued and full-length gene (encoding the entire protein) patents have not been used very restrictively, the overall patent climate has nonetheless become quite menacing. Problems in gaining

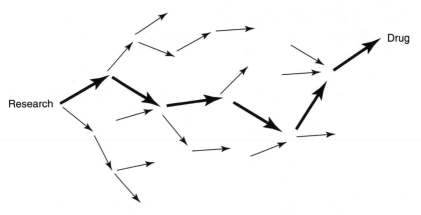

FIGURE A.2 Pathways to new medicine: Where does basic research start and applied research begin?

access to new enabling technologies, the tools used in much groundbreaking research, have become a source of great tension among publicly funded researchers. The NIH director Harold Varmus formed an advisory committee to investigate the issue. Its 1998 report found that universities and government institutions must often endure lengthy negotiations, pay hefty fees, assume certain legal risks (through indemnification clauses), and give up a certain rights to follow-on inventions or discoveries (via *reach-through* clauses) just to gain access to new technologies, even technologies developed at other universities or at government institutions.[10] The committee recommended a number of actions that could greatly facilitate the transfer of intellectual property. The NIH does not have authority over patent matters and the committee did not make recommendations regarding patent procedures. However, they were sufficiently moved to report "considerable dissatisfaction with the current operation of the patent system in biomedical research from many quarters, suggesting that there may be considerable room for improvement."

Dissatisfaction is duly noted, but is scientific progress being hindered by patent laws? Could so many new research tools be developed without the incentive that patent policies create? Consider a technology that is at the heart of the Human Genome

Project—DNA sequencing machines. After Leroy Hood and colleagues invented the first automated sequencing machine at California Institute of Technology (Caltech), no less than 19 companies turned down Hood's request to help advance this technology and make it widely available. The NIH itself turned down several requests to support further development of the technology. Finally, the company Applied Biosystems took up the cause, but only after obtaining an exclusive license to the patent-protected technology. Applied Biosystems then invested over $70 million and produced a highly efficient and widely popular DNA sequencing machine. Applied Biosystems sequencing machines have been used to discover thousands of new genes, and there is little doubt that patent protection helped facilitate the advancement and dissemination of this technology.[11]

Could patents hinder medical progress? Possibly! An NIH advisory committee stated that "virtually every firm that we spoke with believed that restricted access to research tools is impeding the rapid advance of research and that the problem is getting worse."

Medical progress, and in particular the development of new drugs, is dependent upon commercial enterprises. As more and more enabling technologies and materials, including genes, become patented, and as more and more patent holders seek "reach-through" rights, the ability to navigate and negotiate through this tangle may become onerous. The ability to produce medical advances is now more likely than ever to require a complex web of materials and technologies. Michael Heller and Rebecca Eisenberg of the University of Michigan Law School articulated the problem well in a 1998 article in the journal *Science*, where they wrote that:

> . . . a resource is prone to underuse in a "tragedy of the anticommons" when multiple owners each have a right to exclude others from a scarce resource and no one has effective privilege of use. In theory, in a world of costless transactions, people could always avoid commons or anticommons tragedies by trading their rights. In practice, however, avoiding tragedy requires overcoming transaction costs, strategic behaviors, and cognitive biases of

participants, with success more likely within close-knit communities than among hostile strangers. Once an anticommons emerges, collecting rights into usable private property is often brutal and slow.[12]

One the main theses of this book is that we are being drawn towards molecular medicine (and perhaps something else) much like a heavy object being drawn into a black hole. If this is true, then a closer-knit community will form, one way or another. Businesses have ways of dealing with the anticommons problem (see Chapter 15).

Can the patent process be changed? Absolutely! The legal code is not chiseled in stone. Like many other codes, it continually evolves. Patent laws and procedures change through court decisions following legal challenges and through legislation. In light of the accelerating pace of technological changes and the apparent dissatisfaction in the patent process, it seems that revolutionary change, systemic and procedural revision, could also come to patent law. Incremental changes are more likely, however. Rules could be changed so as to better ensure that allowed claims are supported by the specification and are not overly broad.

Can a compromise be reached between those who oppose animal, cell, protein, and gene patents and those who wish to commercialize the ideas and materials disclosed in these patents? "We believe that humans and animals are creations of God, not humans, and as such should not be patented as human inventions," announced the Joint Appeal Against Human and Animal Patenting at a press conference in May of 1995.[13] The group is a coalition of leaders and representatives from many different religious faiths that oppose the patenting of genes and cells, as well as whole organisms. In response, Lisa Raines, a vice president of Genzyme Corporation, echoed a sentiment held by most gene and animal researchers, "Our goals are not to play God; they are to play doctor." Many people feel that the ability and desire to probe the mysteries of life and intervene in disease processes are God-given. The Joint Appeal Against Human and Animal Patenting acknowledged that "the new techniques in genetic engineering offer exciting possibilities for the curing of disease and for helping to preserve nature's diversity." Thus, there does not appear to be a great deal of conflict, and perhaps some sort of compromise may be possible. The question is how to utilize our new techniques in ways that improve human health and that preserve human dignity, integrity, and humility. The current patent system does a good job of helping to propel medical advances, but if it degrades core human values, then an alternative must be found. Both scientists and members of the clergy have been examining alternatives to gene patents that may satisfy both parties.[14]

Glossary

A, C, T, G   The four nitrogen-containing molecules that make up DNA. A fifth molecule, U, substitutes for T in RNA. The names of these molecules are as follows: A–adenine, C–cytosine, T–thymine, G–guanine, and U–uracil.

allele   A particular gene variant.

amino acid   A group of 20 naturally occurring chemical compounds that are the building blocks of proteins.

antibody   A defensive protein secreted by B-lymphocytes (a type of white blood cell) that is capable of recognizing and binding to molecules that are foreign to the organism.

base pair   A pair of nucleotides on opposing strands of a DNA molecule. As pair with Ts, and Cs pair with Gs.

bioinformatics   The scientific discipline concerned with the analysis of the information contained in or transmitted by DNA, RNA, protein, or other biological molecules.

cDNA   Complementary DNA. DNA made in laboratory using mRNA as a template.

cDNA library   A collection of cDNAs, usually generated from one particular tissue or cell source.

chromosome   A very large and usually highly structured DNA molecule and bound proteins, capable of replicating and then segregating into daughter cells.

clone   A collection of identical organisms derived from a single parent, or a collection of identical DNA molecules derived from the same DNA template. As a verb, *to clone* means to make an exact copy.

DNA   Deoxyribonucleic acid, a two-stranded double helical molecule composed of a series of aligned nucleic acids.

EST   *Expressed Sequence Tag*. The sequence of a small portion of a gene. ESTs are usually sufficient to identify (or tag) a gene and may provide enough information to indicate the function of a gene.

evolution   The continuous genetic adaptation of organisms or populations through mutation, random drift, and natural selection.

functional genomics   Efforts to determine the function of a large number of genes, usually using automated high-throughput processes.

gene   An inherited packet of biological information. The sequence of nucleic acids that encodes a protein or RNA molecule, or an observable trait. The information for a protein or RNA molecule, or an observable trait.

gene chip   See *microarray*.

gene expression   The multistep process in which RNA and proteins are synthesized. *Gene expression levels* refer to the relative quantity of RNA or protein.

gene variant   One of several forms of a gene, where the underlying sequence varies. Also known as an *allele*.

genetic code   The rules that dictate how nucleotide triplets in DNA correspond to the amino acid sequence in proteins.

genome   A term created in 1920 by combining the words GENes and chromosOMEs. The genome is a complete set of chromosomes and their genes.

genomics   Word coined in 1986 by Thomas Roderick to describe the scientific discipline of mapping, sequencing, and analyzing genomes. It is now most commonly used to describe the scientific discipline and industry involved in accessing and analyzing thousands of different biological molecules at a time.

genotype   The genetic constitution of an organism. It usually refers to the composition of alleles in one or more genes of interest.

homolog   A sequence that is very similar to another, probably arising from a common ancestral sequence.

hybridization   The process of joining two complementary strands of nucleic acids.

linkage   The association of genes on the same chromosome.

microarray   A two-dimensional surface upon which thousands of molecules are arranged in discrete positions. Used to simultaneously assay thousands of distinct molecules that interact with the arrayed molecules.

mRNA *Messenger RNA.* RNA that conveys information directing the synthesis of proteins.

mutation A change in DNA sequence.

PCR *Polymerase Chain Reaction,* a technique for amplifying short stretches of DNA, by which millions copies of a sequence may be derived from as few as one copy.

pharmacogenetics The study of how gene variations affect the safety and efficacy of drugs and drug candidates.

pharmacogenomics The study of drug and candidate drug actions using genomewide assays, such as those that measure large-scale gene expression changes.

phenotype Appearance.

polymorphism A variation in sequence.

protein A class of molecules consisting of one or more polypeptides. Polypeptides are chains of amino acids.

protein expression The process by which proteins are synthesized. *Protein expression levels* refer to the relative quantity of protein in a tissue or cell.

proteome The set of all proteins of a particular cell, tissue, or organism. Or, as scientists Marc Wilkins and Keith Williams have stated, "the set of PROTEins encoded by the genOME."

proteomics The study of a cell, tissue, or organism's entire set of proteins, their expression level, location, state of activity, and so forth.

RNA *Ribonucleic acid.* A single strand of nucleotides derived from a DNA template, which may carry information for making proteins or act in catalyzing certain cellular reactions.

sequence A series of symbols. With DNA and RNA it refers to a series of base pairs or nucleotides. With proteins it refers to a series of amino acids in the peptide chains.

single nucleotide polymorphism (SNP, pronounced "snip") A particular class of DNA sequence variations found among individuals of the same species.

template A mold or cast from which multiple copies can be made. A strand of DNA from which multiple copies of RNA or DNA strands can be made.

transcript The product of transcription, an RNA or mRNA molecule.

transcript image   A list or image depicting the relative abundance of transcripts within a sample of living cells.

transcription   The process of producing a chain of ribonucleotides (an RNA molecule or transcript) whose sequence is derived from the nucleotide sequence of a DNA molecule.

transduce   To convert energy or information from one form into another, particularly into an electrical form. (Transduce also has another meaning that is used in microbiology, but which is not applicable here.)

translation   The process of producing a chain of amino acids whose sequence is derived from the nucleotide sequence of a mRNA molecule.

two-dimensional gel electrophoresis   A technique for separating a group of proteins in which proteins are spread apart in a gel by the application of an electrical current in one direction and a chemical gradient in the orthogonal direction.

Notes

CHAPTER 1

1. Underlying genetic variations in tumor tissue may also be reflected in the cells' overall protein composition, as is discussed in later chapters.

2. Anderson, L., Hofmann, J., Gemmel, A., and Taylor, J., "Global Approaches to Quantitative Analysis of Gene-Expression Patterns Observed by the Use of Two-Dimensional Gel Electrophoresis," *Clinical Chemistry* **30**:2031–2036 (1984).

3. Anderson, N., and Anderson, L., "The Human Protein Index." *Clinical Chemistry* **28**:739–747 (1982).

4. Anderson, L., Hofmann, J., Gemmel, A., and Taylor, J., "Global Approaches to Quantitative Analysis of Gene-Expression Patterns Observed by the Use of Two-Dimensional Gel Electrophoresis." *Clinical Chemistry* **30**:2031–2036 (1984).

5. Anderson, N., and Anderson, L., "The Human Protein Index." *Clinical Chemistry* **28**:739–747 (1982).

6. Quoted by Nicolas Wade in "The Complete Index of Man," *Science* **211**:33–35 (1981).

7. Humans have 23 pairs of chromosomes, but one differs significantly between males and females; thus 24 distinct chromosomes.

8. Quoted by Nicolas Wade in "The Complete Index of Man," *Science* **211**:33–35 (1981).

9. Ibid.

10. Boyd, M., "National Cancer Institute Drug Discovery and Development." In Frei, E.J. and Freireich, E.J. (eds): *Accomplishments in Oncology, Vol. 1, Cancer Therapy: Where Do We Go From Here?* pp. 68–76 J.B. Lippincott, Philadelphia (1986).

11. Myers, T.G. et al., "A Protein Expression Database for the Molecular Pharmacology of Cancer," *Electrophoresis* **18**:647–653 (1997).

12. Weinstein, J.N., et al., "An Information-Intensive Approach to the Molecular Pharmacology of Cancer," *Science* **275**:343–349 (1997).
13. Ibid.

CHAPTER 2

1. SETI web site: www.seti.org/faq/html.
2. Weaver, W., and Shannon, C.E., *The Mathematical Theory of Communication*, University of Illinois Press, Urbana, Illinois (1949).
3. Loewenstein, W., *The Touchstone of Life*, Oxford University Press, New York (1999).

CHAPTER 3

1. Weiner, J., *Time, Love, Memory: A Great Biologist and His Quest for the Origins of Behavior*, Knopf, New York (1999).
2. "Online Mendelian Inheritance in Man," OMIM. Center for Medical Genetics, Johns Hopkins University (Baltimore, MD) and National Center for Biotechnology Information, National Library of Medicine (Bethesda, MD), 2001. www.ncbi.nlm.nih.gov/Omim/Stats/mimstats.html.
3. Dryja, T.P., "Gene-Based Approach to Human Gene-Phenotype Correlations," *Proc. Nat. Acad. Sci.* **94**:12117–12121 (1997).
4. Donahue, R.P., Bias, W.B., Renwick, J.H., and McKusick, V.A., "Probable Assignment of the Duffy Blood Group Locus to Chromosome 1 In Man." *Proc. Nat. Acad. Sci.* **61**:949–955 (1968).
5. Many positionally cloned genes are listed at: www.ncbi.nlm.nih.gov/Bassett/dbEST/PosiClonNew.html.
6. Reviewed by Capel, B., "Sex In the 90s: SRY and the Switch to the Male Pathway," *Annu. Rev. Physio.* **60**:497–523 (1998).
7. Alton, E.W., Geddes, D.M., Gill, D.R., Higgins, C.F., Hyde, S.C., Innes, J.A., and Porteous, D.J. "Towards Gene Therapy for Cystic Fibrosis: A Clinical Progress Report." *Gene Ther.* **5**:291–292 (1998).
8. Zhang, Y., Proenca, R., Maffei, M., Barone, M., Leopold, L., and Friedman, J.M., "Positional Cloning of the Mouse Obese Gene and its Human Homologue," *Nature* **372**:425–432 (1994).

CHAPTER 4

1. Dawkins, Richard, *Climbing Mount Improbable*, Norton, New York (1996) p. 326.
2. GeneBank statistics: www.ncbi.nlm.nih.gov/Genbank/genbankstats. html.

CHAPTER 5

1. Cook-Deegan, Robert, *Gene Wars*, W.W. Norton & Company, New York (1994), p. 285 (1996 paperback edition).
2. By 1996 MEDLINE was indexing about 400,000 articles a year. Titles, authors, abstracts, keywords, publishers, sequences, and more are all available at the click of a mouse (www.ncbi.nlm.nih.gov/ Entrez/medline.html).
3. www.ebi.ac.uk/sprot/summary.html.
4. Burks, C., Fickett, J.W., Goad, W.B., Kanehisa, M., Lewitter, F.I., Rindone, W.P., Swindell, C.D., Tung, C.S., and Bilofsky, H.S. "The GenBank Nucleic Acid Sequence Database." *Comput. Appl. Biosci.* **1**:225–233 (1985).
5. Curiously, people tend to make large databases anthropomorphic, with an identity, feelings, and emotions. They grow, consume energy, need continual care, and may at times be quite stubborn and uncooperative. Their keepers may develop special relationships to them.
6. Altschul, S.F., Gish, W., Miller, W., Myers, E.W., and Lipman, D.J., "Basic Local Alignment Search Tool," *J. Mol. Biol.* **215**:403–410 (1990).
7. Nakata, K., Karehise, M., and Debis, C., "Prediction of Splice Junctions in mRNA Sequences," *Nucleic Acids Res.* **13**:5327–5340 (1985).

CHAPTER 6

1. Cantor, Charles, *The Code of Codes*, Ed. Daniel Kevles and Leroy Hood, Harvard University Press, Cambridge, Massachusetts (1992).
2. A stretch of DNA is a rather fuzzy concept. Unfortunately, it is often not clear exactly which nucleotides are part of a gene and which are not.
3. Wolffe, A.P., and Matzke, M.A., "Epigenetics: Regulation Through Repression," *Science* **286**:481–486 (1999).

4. Ostrow, R.S., Woods, W.G., Vosika, G.J., and Faras, A.J., "Analysis of the Genetic Complexity and Abundance Classes of Messenger RNA In Human Liver and Leukemic Cells." *Biochim. Biophys. Acta.* **562:** 92–102 (1979).

5. Milner R.J., and Sutcliffe, J.G., "Gene Expression In Rat Brain," *Nucleic Acids Res.* **11:**5497–5520 (1983).

6. Sutcliffe, J.G., "mRNA in the Mammalian Central Nervous System." *Annu. Rev. Neurosci.* **11:**157–198 (1988); Fields, C., Adams, M.D., White, O., and Venter, J.C., "How Many Genes In the Human Genome?" *Nat. Genet.* **7:**345–346 (1994).

7. Ostrow, R.S., Woods, W.G., Vosika, G.J., and Faras, A.J., "Analysis of the Genetic Complexity and Abundance Classes of Messenger RNA in Human Liver and Leukemic Cells," *Biochim. Biophys. Acta.* **562:** 92–102 (1979).

8. Ohno, S., "So Much 'Junk' DNA In Our Genome," in *Evolution of Genetic Systems,* Brookhaven Symposium on Biology (ed. H.H. Smith) p. 366–370, Gordon and Breach, New York, (1972).

9. They are called Alu sequences because their discovery relied on the fact that a sequence within it happens to be cleaved by the Alu enzyme named for the bacterium from which it was derived, *Arthrobacter luteus.*

10. Ohno, S., "So Much 'Junk' DNA In Our Genome," in *Evolution of Genetic Systems,* Brookhaven Symposium on Biology (ed. H.H. Smith) p. 366–370, Gordon and Breach, New York, (1972).

11. Dawkins, Richard, *The Selfish Gene,* Oxford University Press (1976).

12. Sverdlov, E.D. "Retroviruses and Primate Evolution," *Bioessays* **22:**161–171 (2000).

13. Olno, S. and Yome, T., "The Grammatical Rule for All DNA: Junk and Coding Sequences," *Electrophoresis* **12:**103–108 (1991).

14. Chu, W.M., Ballard, R., Carpick, B.W., Williams, B.R., and Schmid, C.W., "Potential Alu Function: Regulation of the Activity of Double-Stranded RNA-Activated Kinase PKR," *Mol. Cell. Biol.* **18:**58–68 (1998).

15. Cook-Deegan, Robert, *Gene Wars,* W.W. Norton and Company, New York (1994; 1996 paperback edition).

16. Weinberg, R.A., "The Case Against Sequencing," *The Scientist* **1:**11 (1987).

17. Cook-Deegan, Robert, *Gene Wars,* W.W. Norton and Company, New York (1994) p. 122 (1996 paperback edition).

18. *The Human Genome Initiative: Issues and Impacts*, p. 96–97.

19. Cook-Deegan, Robert, *Gene Wars*, (1994) p. 111 (1996 paperback edition).

20. Cook-Deegan, Robert, *Gene Wars*, (1994) p. 89 (1996 paperback edition).

21. Genome Corporation Business Plan (1987), courtesy of Walter Gilbert.

22. Gilbert later cofounded Myriad Genetics, a positional cloning and diagnostics company, and continued to study gene evolution at Harvard University.

23. *Code of Codes*, p.27.

24. Gilbert, W., "Towards a Paradigm Shift in Biology," *Nature* **349**:99 (1991).

CHAPTER 7

1. Adams, M., Kelley, J., Gocayne, J., Dubnick, M., Polymeropoulos, M., Xiao, H., Merril, C., Wu, A., Plde, B., Moreno, R., Kerlavage, A., and McCombie, R., "Complementary DNA Sequencing: Expressed Sequence Tags and Human Genome Project," *Science* **252**:1651–1656 (1991).

2. Roberts, L. "Genome Patent Fight Erupts," *Science* **254**:184–188 (1991).

3. Rifkin, Jeremy, "Patent Pending," *Mother Jones* May/June (1998).

4. U.S. Constitution, article I, Section 8, Clause 8, 1789.

5. Roberts, L. "NIH Gene Patents, Round Two," *Science* **255**: 912–913 (1992).

6. "Venter's Venture" (editorial), *Nature* **362**:575–576 (1993).

7. Berselli, Beth, "Gene Split Research Partners Human Genome and TIGR Are Ending Their Marriage of Convenience," *Washington Post*, July 7, 1997.

8. I began working at Incyte in 1996 and often heard Randy Scott declare that through the genomics revolution "the molecular basis of all human diseases will be known within 10 years." I was skeptical. I knew of the power of recombinant DNA technology, which was developed decades earlier, and had conducted research at Stanford with those that had developed it and at Genentech Inc. with those that had first applied it towards improving human health. And I knew that despite the immense advances in molecular biology we

were still very much in the dark as to the moleular underpinnings of most disease. But I also had first-hand knowledge of Silicon Valley's powerful engines of innovation and the can-do spirit that resides there. After working with Incyte two years, I too became convinced of the power and potential of genomics. The evidence of a fundamental shift in the biomedical sciences was compelling, and it helped inspire me to write this book.

CHAPTER 8

1. Hoke, F., "Limited Access To cDNA Database Has Drug Manufacturer Up In Arms," *The Scientist* **8**: 1, 8–9 (1994).
2. Caskey, C.T., et al., "Merck, SmithKline and patents," *Nature* **381:** 360 (1996); Nowak, R., "Merck Releases First 'Gene Index' Sequences," *Nature* **373:**549 (1995); "Merck Hires Top Academic Geneticist," *Science* **266:**538 (1994) Macilwain C. Caskey was recruited to head Merck's genome efforts. *Nature*, Oct 27, 1994.
3. Merck and Company press release, Feb 10, 1995.
4. From Merck and Company home page: www.merck.com
5. Boguski, M.S., "The Turning Point in Genome Research." *Trends Biochem. Sci.* **20:**295–296 (1995).
6. Benson, D.A., Boguski, M.S., Lipman, D.J., Ostell, J., and Ouellette, B.F., "GenBank," *Nucleic Acids Res.* **26:**1–7 (1995).
7. Levy-Lahad, E., Wasco, W., Poorkaj, P., Romano, D.M., Oshima, J., Pettingell, W.H., Yu, C.E., Jondro, P.D., Schmidt, S.D., Wang, K, et al., "Candidate Gene for the Chromosome 1 Familial Alzheimer's Disease Locus," *Science* **269:**973–977 (1995).
8. Wiley, S.R., Schooley, K., Smolak, P.J., Din, W.S., Huang, C.P., Nicholl, J.K., Sutherland, G.R., Smith, T.D., Rauch, C., Smith, C.A., et al., "Identification and Characterization of a New Member of the TNF Family That Induces Apoptosis," *Immunity* **3:**673–682 (1995).
9. www.ncbi.nlm.nih.gov/dbEST/CancerGene.html.
10. www.ncbi.nlm.nih.gov/Bassett/dbEST/PosiClonNew.html.
11. www.tigr.org/tdb/hgi/hgi.html.
12. Pennisi, E. "And the Gene Number is . . . ?," *Science* **288:**1146–1147 (2000).
13. Bonaldo, M.F., Lennon, G., and Soares, M.B., "Normalization and Subtraction: Two Approaches to Facilitate Gene Discovery," *Genome Res.* **6:**791–806 (1996).

CHAPTER 9

1. www.hgsi.com/news/press/97-12-18_MPIF1.html.

2. Haseltine, W.A. "The Power of Genomics to Transform the Biotechnology Industry," *Nat. Biotech.* **16 Suppl:**25–27 (1998).

3. Strings of As, Ts, Cs, and Gs had always been a staple of gene sequence articles, but interesting as they were, *H. influenzae*'s 1.8 million letters would not grace the pages of *Science* magazine. Colorful foldout gene maps highlighted the text, but readers were directed to GenBank to view the fine details.

4. Nowak, Rachel, "Bacterial Genome Sequence Bagged," *Science* **269:**469 (1995).

5. Ibid.

6. www.ornl.gov/hgmis/faq/faqs1.html#q4.

7. Stewart, E.A., McKusick, K.B., Aggarwal, A. et al., "An STS-Based Radiation Hybrid Map of the Human Genome," *Genome Research* **7:**422–433 (1997).

8. Schuler, G.D., Boguski, M.S., Stewart, E.A., et al., *Science* 1996; **274:**540–546).

9. Rowen, L., Mahairas, G., and Hood, L., "Sequencing the Human Genome," *Science* **278:**605–607 (1997).

10. Ibid.

11. Weber, J.L., and Myers, E.W., "Human Whole-Genome Shotgun Sequencing," *Genome Res.* **7:**401–409 (1997).

12. Ibid.

13. Venter, J.C., Adams, M.D., Sutton, G.G., Kerlavage, A.R., Smith, H.O., and Hunkapiller, M. "Shotgun Sequencing of the Human Genome," *Science* **280:**1540–1542 (1998).

14. Ibid.

15. *New Scientist*, May 23, 1998 and *Science* article.

16. www.wellcome.ac.uk/wellcomegraphic/a6/g3index.html.

17. Testimony of Maynard Olson before the House Committee on Science, Subcommittee on Energy and Environment June 17th, 1998.

18. Collins, F., Patrinos, A., Jordan, E., Chakravart, A., Gesteland, R., and Walters, L., "New Goals for the U.S. Human Genome Project: 1998–2003," *Science,* **282:**682 (1998).

19. Ibid.

20. *Jedermann sein eigner Fussball* (1919), a journal edited by George Grosz and Franz Jung, Berlin.

21. Good intentions (greater human knowledge, and better human health), have been virtually universal in this endeavor. There has been no evidence of harmful intents, nor has there been evidence that knowledge of human genes is being used to harm people. However, there certainly is the potential to use DNA sequence information in malevolent ways, for example to unfairly discriminate against particular people or to construct biological weapons. Thus, there is every reason for people to be wary.

22. The race is a success in that the sequencing and initial gene discovery and characterization goals are being realized. The "success" of a new plane of self-awareness and new control over our physical being remains to be seen.

CHAPTER 10

1. Garcia Marquez, G. *One Hundred Years of Solitude,* Avon Books, New York (1970).

2. Qadota, H., Anraku, Y., Botstein, D., and Ohya, Y., "Conditional Lethality of a Yeast Strain Expressing Human RHOA in Place of RHO1." *Proc. Natl. Acad. Sci.* **91**:9317–9321 (1994).

3. Makalowski, W., Zhang, J., and Boguski, M.S. "Comparative Analysis of 1196 Orthologous Mouse and Human Full-length mRNA and Protein Sequences," *Genome Research* **6**:846–857 (1996).

4. Figuera, L.E., Pandolfo, M., Dunne, P.W., Cantu, J.M., Patel, P.I., "Mapping of the Congenital Generalized Hypertrichosis Locus to Chromosome Xq 24-q 27.1," *Nat. Genet.* **10**:202–207 (1995); Neri, G., Gurrieri, F., Zanni, G., and Lin, A., "Clinical and Molecular Aspects of the Simpson-Golabi-Behmel Syndrome," *Am. J. Med. Genet.* **79**:279–283 (1998); Tsukahara, M., Uchida, M., Uchino, S., Fujisawa, R., Kamei, T., and Itoh, T., "Male to Male Transmission of Supernumerary Nipples," *Am. J. Med. Genet.* **69**:194–195 (1997).

5. The words "allele" and "polymorphism" are typically used to describe sequence variants that are firmly established in a population. The word "mutation" is typically used to describe relatively recent, rarer, or disease-associated sequence variations. The meanings overlap and are somewhat arbitrary.

6. Nathans, J., "Molecular Genetics of Human Visual Pigments," *Annu. Rev. Genet.* **26**:403–424 (1992).

7. Diamandis, E.P., Yousef, G.M., Luo, I., Magklara, I., and Obiezu, C.V. "The New Human Kallikrein Gene Family: Implications in Carcinogenesis," *Trends Endocrinol Metab.* **11**:54–60 (2000).

8. Chao, D.T., and Korsmeyer, S.J., "BCL-2 Family: Regulators of Cell Death," *Annu. Rev. Immunol.* **16**:395–419 (1998).

9. Bird, A., and Tweedie, S., "Transcriptional Noise and the Evolution of Gene Number," *Philos. Trans. R. Soc. Lond. B. Biol. Sci.* **349**: 249–253 (1995).

CHAPTER 11

1. White House press release June 26, 2000 www.whitehouse.gov/WH/New/html/genome-20000626.html.

2. Roundtable Forum "The Human Genome Initiative: Issues and Impacts" Office of Health and Environmental Research. U.S. Department of Energy, Washington DC 20545 (1987).

3. Nei, M., and Roychoudhury, A.K., "Gene Differences Between Caucasian, Negro, and Japanese Populations," *Science.* **177**:434–436 (1972).

4. Recent studies indicate that variations in the melanocortin-stimulating hormone receptor gene underlie much of the observed variation in human skin and hair color. See Rana, B.K. *Genetics* 151: 1547 and Schioth, H.B., et al, *Biochem. Biophys. Res. Commun.* **260**: 488 (1999).

5. Nishikimi, M., Fukuyama, R., Minoshima, S., Shimizu, N., and Yagi, K., "Cloning and Chromosomal Mapping of the Human Nonfunctional Gene for L-Gulono-Gamma-Lactone Oxidase, the Enzyme for L-Ascorbic Acid Biosynthesis Missing In Man." *J. Biol. Chem.* **269**:13685–13688 (1994).

6. Samson, M., Libert, F., Doranz, B.J., Rucker, J., Liesnard, C., Farber, C.M., Saragosti, S., Lapoumeroulie, C., Cognaux, J., Forceille, C., Muyldermans, G., Verhofstede, C., Burtonboy, G., Georges, M., Imai, T., Rana, S., Yi, Y., Smyth, R.J., Collman, R.G., Doms, R.W., Vassart, G., and Parmentier, M. "Resistance to HIV-1 Infection In Caucasian Individuals Bearing Mutant Alleles of the CCR-5 Chemokine Receptor Gene," *Nature* **382**:722–725 (1996); and Liu, R., Paxton, W.A., Choe, S., Ceradini, D., Martin, S.R., Horuk, R., MacDonald, M.E., Stuhlmann, H., Koup, R.A., and Landau, N.R., "Homozygous Defect in HIV-1 Co-receptor Accounts for Resis-

tance of Some Multiply-Exposed Individuals to HIV-1 Infection," *Cell* **86**:367–377 (1996).

CHAPTER 12

1. Miki, Y., Swensen, J., Shattuck-Eidens, D., Futreal, P.A., Harshman, K., Tavtigian, S., Liu, Q., Cochran, C., Bennett, L.M., Ding, W., Bell, R., Rosenthal, and 33 others, "A Strong Candidate for the Breast Cancer and Ovarian Cancer Susceptibility Gene BRCA1." *Science* **266**:66–71 (1994).

2. Wooster, R., Bignell, G., Lancaster, J., Swift, S., Seal, S., Mangion, J., Collins, N., Gregory, S., Gumbs, C., and Micklem, G., "Identification of the Breast Cancer Susceptibility Gene BRCA2," *Nature* **378**: 789–792 (1995).

3. Phelan, C.M., et al., "Ovarian Risk in BRCA1 Carriers is Modified by the HRAS1 Variable Number of Tandem Repeat (VNTR) Locus," *Nature Genet.* **12**:309–311; and Easton, D.F., Steele, L., Fields, P., Ormiston, W., Averill, D., Daly, P.A., McManus, R., Neuhausen, S.L., Ford, D., Wooster, R., Cannon-Albright, L.A., Stratton, M.R., and Goldgar, D.E., "Cancer Risks in Two Large Breast Cancer Families Linked to BRCA2 on Chromosome 13q12-13," *Am. J. Hum. Genet.* **61**:120–128 (1997).

4. Warmuth, M.A., Sutton, L.M., and Winer, E.P. "A Review of Hereditary Breast Cancer: From Screening to Risk Factor Modification," *Am. J. Med.* **102**:407–415 (1997).

5. Sellers, T.A., "Genetic Factors in the Pathogenesis of Breast Cancer: Their Role and Relative Importance," *J Nutrit* **127**:929S–932S (1997).

6. Brugarolas, J. and Jacks, T. "Double Indemnity: p53, BRCA and Cancer. p53 Mutation Partially Rescues Developmental Arrest in Brca1 and Brca2 Null Mice, Suggesting a Role for Familial Breast Cancer Genes in DNA Damage Repair," *Nat. Med.* **3**:721–722 (1997).

7. Prescott, C.A. and Gottesman, I.I., "Genetically Mediated Vulnerability to Schizophrenia," *Psychiatr. Clin. North Am.* **16**:245–267 (1993).

8. Moises, H.W., et al., "An International Two-Stage Genome-Wide Search for Schizophrenia Susceptibility Genes," *Nature Genet.* **11**: 321–324 (1995).

9. Moldin, S.O., and Gottesman, I.I., "At Issue: Genes, Experience, and Chance In Schizophrenia—Positioning for the 21st Century," *Schizophr. Bull.* **23**:547–561 (1997).

10. Niinaka, Y., Paku, S., Haga, A., Watanabe, H. and Raz, A., "Expression and Secretion of Neuroleukin/phosphohexase isomerase/maturation factor as autocrine motility factor by tumor cells." *Cancer Res.* **58**:2667–2674 (1998).

11. Levins, Richard, and Lewontin, Richard, *The Dialectical Biologist,* Harvard University Press, Cambridge, Massachusetts, p. vii (1985).

12. Strohman, R.C. "The Coming Kuhnian Revolution in Biology," *Nat Biotech* **15**:194–200 (1997).

13. Kuhn, Thomas, *The Structure of Scientific Revolutions.* University of Chicago Press, Chicago (1962).

CHAPTER 13

1. Crick conceived the central dogma to describe only sequence information flow. However, it has unfortunately come to be interpreted more broadly to imply that all information flows linearly from DNA to trait.

2. Schrödinger, Erwin, *What is Life?* (p. 21 of 1998 paperback version) Cambridge University Press (1944).

3. Lukashin, A. and Borodovsky, M., "GeneMark.hmm: New Solutions for Gene Finding," *Nucleic Acids Res.* **26**:1107–1115 (1998).

4. Berg, J.M., "Proposed Structure for the Zinc-Binding Domains from Transcription Factor IIIA and Related Proteins," *Proc. Natl. Acad. Sci.* **85**:99–102; and Parraga, G., Horvath, S.J., Eisen, A., Taylor, W.E., Hood, L., Young, E.T., and Klevit, R.E., "Zinc-Dependent Structure of a Single-Finger Domain of Yeast ADR1," *Science* **241**:1489–1492 (1988).

5. Clarke, N.D., and Berg, J.M., "Zinc Fingers in Caenorhabditis Elegans: Finding Families and Probing Pathways," *Science.* **282**:2018–2022 (1998).

6. Collado-Vides, J., "Towards a Unified Grammatical Model of Sigma 70 and Sigma 54 Bacterial Promoters," *Biochimie* **78**:351–363 (1996).

7. Jacob, F. and Monod, J. "Genetic Regulatory Mechanisms in the Synthesis of Proteins." *Journal of Molecular Biology* **3**:318–356 (1961).

8. Ptashne's work is carefully documented in his book *A Genetic Switch:* Gene Control and Phage Lamda Cell Press, Cambridge MA (1987).

9. McAdams, H. and Shapiro, L., "Circuit Simulation of Genetic Networks," *Science* **269**:650–656 (1995).

10. Kauffman, Stuart, *The Origins of Order: Self-Organization and Selection in Evolution*, Oxford Univ. Press (1993).

11. Kauffman, Stuart, *At Home in the Universe*, Oxford University Press (1995).

12. Anderson, L., Hofmann, J., Gemmel, A., and Taylor, J. "Global Approaches to Quantitative Analysis of Gene-Expression Patterns Observed by the Use of Two-Dimensional Gel Electrophoresis." *Clinical Chemistry* **30**:2031–2036 (1984).

13. *Biocomputing '98* World Scientific Edited by Russ Altman et al. (1998).

CHAPTER 14

1. Augenlicht, L.H., and Kobrin, D., "Cloning and Screening of Sequences Expressed in a Mouse Colon Tumor," *Cancer Res.* **42**: 1088–1093 (1982).

2. Noncommercial microarray development occurred elsewhere. For example, Roger Ekins of the University College in London began arraying oligonucleotides for hybridizati n studies as early as 1986. A few year later there were several oth engineering teams working on what was sometimes referred to ; a multiplexed biosensor at that time. Nonetheless, microarray c velopment work in Silicon Valley appears to have had a far greate impact on biomedicine than these earlier efforts.

3. State of the Union address Jan. 1998 w w.pub.whitehouse.gov/WH/ Publications/html/publications.html

4. Ibid.

5. Stipp, David, "Gene Chip Breakthrough," *Fortune* 3/31/97, p.56.

6. Ibid.

7. Fodor, S.P., Read, J.L., Pirrung, M.C., Stryer, L., Lu, A.T., and Solas, D., "Light-Directed, Spatially Addressable Parallel Chemical Synthesis," *Science* **251**:767–773 (1991).

8. Lockhart, D.J., Dong, H., Byrne, M.C., Follettie, M.T., Gallo, M.V., Chee, M.S., Mittmann, M., Wang, C., Kobayashi, M., Horton, H., and Brown, E.L., "Expression Monitoring by Hybridization to High-Density Oligonucleotide Arrays," *Nat Biotechnol.* **14:**1675–1680 (1996).

9. Audic, S., and Claverie, J.M., "The Significance of Digital Gene Expression Profiles," *Genome Res.* **7:**986–995 (1997).

10. Okubo, K., Hori, N., Matoba, R., Niiyama, T., Fukushima, A., Kojima, Y., and Matsubara, K., "Large Scale cDNA Sequencing for Analysis of Quantitative and Qualitative Aspects of Gene Expression," *Nat. Genet* **2:**173–179 (1992).

11. Walker, M.G., Volkmuth, W., Sprinzak, E., Hodgson, D., and Klingler, T., "Prediction of Gene Function by Genome-Scale Expression Analysis: Prostate Cancer-Associated Genes," *Genome Res.* **9:** 1198–1203 (1999).

CHAPTER 15

1. Biotechnology Industry Organization (BIO) Web site www.bio.org.

2. Schoemaker, H.P., and Shoemaker, A.F., "The Three Pillars of Bioentrepreneurship," *Nature Biotechnology* **16:**13–15 (1998).

3. Recombinant Capital, www.recap.com.

4. According to Jean Delage, a Bay Area venture capitalist, in the *Business Journal of San Jose* 12/7/98.

5. Stryer, Lubert, *Biochemistry,* W.H. Freeman & Co. San Francisco (1981).

6. Rifkin, Jeremy, *The Biotech Century,* pp. 233–234, Penguin Putnam, New York (1998).

CHAPTER 16

1. Marshall, E. "Drug Firms to Create Public Database of Genetic Mutations" *Science* **284:**406 (1999) and " 'Playing Chicken' over Gene Markers." *Science* **278:**2046–2048. (1997).

2. Sherry, S.T., Ward, M., and Sirotkin, K., "dbSNP-database for single nucleotide polymorphisms and other classes of minor genetic variation." *Genome Research* **9:**677–679 (1999); www.ncbi.nlm.nih.gov/ SNP/.

3. Goldman, Bruce, "SNPs: Patent la différence," Recap's *Signals* magazine www.recap.com/signalsmag.nsf 5/19/98

4. Arthur Holden SNP Consortium CEO in *Signals* magazine article: Goldman, Bruce "Industrial-Strength Genomics" 8/24/99. www.recap.com/signalsmag.nsf.

5. Russo, Eugene, and Smaglik, Paul, *The Scientist* **13**:1 (1999).

6. Quoted in Hodgson, John "Analysts, firm pour cold water on SNP Consortium" *Nature Biotechnology* **17**:526 (1999).

7. Wang, D.G., Fan, J.B., Siao, C.J., et. al, "Large-Scale Identification, Mapping, and Genotyping of Single-Nucleotide Polymorphisms in the Human Genome" *Science* **280**:1077–1082 (1998); and Cargill, M., Altshuler, D., Ireland, J., et al., "Characterization of Single-Nucleotide Polymorphisms in Coding Regions of the Human Genes." *Nature Genetics* **22**:233–240 (1999).

8. SNPs in the coding region of genes are sometimes called cSNPs. However, the definition of cSNP is rather loose. Some people use the term to refer to only the subset of coding sequence SNPs that alter the amino acid sequence of the encoded protein.

9. Cargill, M., Altshuler, D., Ireland, J., et al., "Characterization of Single-Nucleotide Polymorphisms in Coding Regions of Human Genes," *Nature Genetics* **22**:231–238 (1999); and Halushka, M.K., Fan, J.B., Bentley, K., et al., "Patterns of Single-Nucleotide Polymorphisms in candidate genes for blood-pressure homeostasis." *Nature Genetics* **22**:239–247. (1999).

10. Although by definition the nonfunctional SNPs do not effect phenotype, they could have a function in a future generation. For example, this variation could provide additional stepping stones for evolutionary change than would not otherwise be available.

11. Roses, A.D., "Pharmacogenetics and the Practice of Medicine," *Nature* **405**:857–865 (2000).

12. Risch, N., and Merikangas, K., "The Future of Genetic Studies of Complex Human Diseases," *Science* **273**:1516–1517 (1996).

13. Dryja, T.P. "Gene-Based Approach to Human Gene-Phenotype Correlations," *Proc. Natl. Acad. Sci.* **94**:12117–12121 (1997).

14. Buetow, K.H., Edmonson, M.N., and Cassidy, A.B., "Reliable Identification of Large Numbers of Candidate SNPs from Public EST Data." *Nat. Genet.* **21**:323–325 (1999).

15. Chee, M., Yang, R., Hubbell, E., Berno, A., Huang, X.C., Stern, D., Winkler, J., Lockhart, D.J., Morris, M.S., Fodor, S.P., "Accessing Genetic Information with High-density DNA Arrays," *Science*. **274**: 610–614 (1996).

16. Wang, D.G., Fan, J.B., Siao, C.J., Berno, A., Young, P., Sapolsky, R., Ghandour, G., Perkins, N., Winchester, E., Spencer, J., Kruglyak, L., Stein, L., Hsie, L., Topaloglou, T., Hubbell, E., Robinson, E., Mittmann, M., Morris, M.S., Shen, N., Kilburn, D., Rioux, J., Nusbaum, C., Rozen, S., Hudson, T.J., Lander, E.S., et al., "Large-Scale Identification, Mapping, and Genotyping of Single-Nucleotide Polymorphisms in the Human Genome," *Science*. **280**:1077–1082 (1998).

17. Kruglyak, L., "Prospects for Whole-Genome Linkage Disequilibrium Mapping of Common Disease Genes," *Nat. Genet.* **22**: 139–144 (1999).

18. Wang, D.G., Fan, J.B., Siao, C.J., Berno, A., Young, P., Sapolsky, R., Ghandour, G., Perkins, N., Winchester, E., Spencer, J., Kruglyak, L., Stein, L., Hsie, L., Topaloglou, T., Hubbell, E., Robinson, E., Mittmann, M., Morris, M.S., Shen, N., Kilburn, D., Rioux, J., Nusbaum, C., Rozen, S., Hudson, T.J., Lander, E.S., et al., "Large-Scale Identification, Mapping, and Genotyping of Single-Nucleotide Polymorphisms in the Human Genome" *Science*. **280**:1077–1082 (1998).

19. Long, A.D., and Langley, C.H., "The Power of Association Studies to Detect the Contribution of Candidate Genetic Loci to Variation in Complex Traits." *Genome Res.* **9**:720–731 (1999).

20. Kruglyak, L., "Prospects for Whole-Genome Linkage Disequilibrium Mapping of Common Disease Genes," *Nat. Genet.* **22**:139–144 (1999).

21. Scott, W.K., Pericak-Vance, M.A., Haines, J.L., Bell, D.A., Taylor, J.A., Long, A.D., Grote, M.N., Langley, C.H., Müller-Myhsok, B., Abel, L., Risch, N., and Merikangas, K., "Genetic Analysis of Complex Diseases." *Science* **275**:1327–1330 (1997).

CHAPTER 17

1. This is known as the Euclidean distance measure. The formula is $D = \sqrt{\Sigma_{i=l}^{N} (x_i - y_i)^2}$

2. Ridker, P.M., Hennekens, C.H., Buring, J.E., Rifai, N., "C-Reactive Protein and Other Markers of Inflammation in the Prediction of

Cardiovascular Disease in Women." *N. Engl. J. Med.* **342**:836–843 (2000).

3. Augenlicht, L.H., Wahrman, M.Z., Halsey, H., Anderson, L., Taylor, J., Lipkin, M., "Expression of Cloned Sequences in Biopsies of Human Colonic Tissue and in Colonic Carcinoma Cells Induced to Differentiate in Vitro." *Cancer Res.* **47**:6017–6021; and Augenlicht, L.H., Taylor, J., Anderson, L., Lipkin, M., "Patterns of Gene Expression that Characterize the Colonic Mucosa In Patients at Genetic Risk for Colonic Cancer." *Proc. Natl. Acad. Sci.* **88**:3286–3289 (1991).

4. Wodicka, L., Dong, H., Mittmann, M., Ho, M.H., and Lockhart, D.J. "Genome-Wide Expression Monitoring in Saccharomyces Cerevisiae." *Nat. Biotechnol.* **15**:1359–1367 (1997).

5. Golub, T.R., Slonim, D.K., Tamayo, P., Huard, C., Gaasenbeek, M., Mesirov, J.P., Coller, H., Loh, M.L., Downing, J.R., Caligiuri, M.A., Bloomfield, C.D., Lander, E.S., "Molecular Classification of Cancer: Class Discovery and Class Prediction by Gene Expression Monitoring," *Science* **286**:531–537 (1999).

6. Alizadeh, A.A., Eisen, M.B., Davis, R.E., Ma, C., Lossos, I.S., Rosenwald, A., Boldrick, J.C., Sabet, H., Tran, T., Yu, X., Powell, J.I., Yang, L., Marti, G.E., Moore, T., Hudson, J. Jr, Lu, L., Lewis, D.B., Tibshirani, R., Sherlock, G., Chan, W.C., Greiner, T.C., Weisenburger, D.D., Armitage, J.O., Warnke, R., Staudt, L.M., et al., "Distinct Types of Diffuse Large B-Cell Lymphoma Identified by Gene Expression Profiling," *Nature* **403**:503–511 (2000).

7. Wen, X., Fuhrman, S., Michaels, G.S., Carr, D.B., Smith, S., Barker, J.L., and Somogyi, R. "Large-Scale Temporal Gene Expression Mapping of the Central Nervous System Development," *Proc. Natl. Acad. Sci. U.S.A.* **95**:334–339 (1998).

8. Eisen, M.B., Spellman, P.T., Brown, P.O., and Botstein, D., "Cluster Analysis and Display of Genome-Wide Expression Patterns," *Proc. Natl. Acad. Sci.* **95**:14863–14868 (1998).

9. Expression clustering is probably sequence based, reflecting similar upstream regulatory regions.

10. Scherf, U., Ross, D.T., Waltham, M., Smith, L.H., Lee, J.K., Tanabe, L., Kohn, K.W., Reinhold, W.C., Myers, T.G., Andrews, D.T., Scudiero, D.A., Eisen, M.B., Sausville, E.A., Pommier, Y., Botstein, D., Brown, P.O., Weinstein, J.N., "A Gene Expression Database for the

Molecular Pharmacology of Cancer," *Nat. Genet.* **24**:236–244 (2000).

11. Ibid.

12. Fayyad, Usama, and Piatesky-Shapiro, Gregory, *Advances in Knowledge Discovery and Data Mining.* Edited by Usama M. Fayyad, Gregory Piatetsky-Shapiro, Padhr Smyth, and Ramasamy Uthurusamy, p.4, MIT Press (1996).

CHAPTER 18

1. Prepared Statement of Craig Venter, Ph.D., President and Chief Scientific Officer Celera Genomics, Before the Subcommittee on Energy and Environment U.S. House of Representatives Committee on Science, April, 6, 2000. Available on Celera's Web site: www.celera.com.

2. Ibid.

3. Article by Tim Friend in *USA Today,* page 7A 3/13/00.

4. Ibid

5. Marshall, E., "Patent on HIV Receptor Provokes an Outcry." *Science* **287**:1375 (2000).

6. Human Genome Sciences press release February 16, 2000, available at www.hgsi.com.

7. Abate, Tom "Call it the Gene Rush—Patent Stakes Run High." *San Francisco Chronicle* April 25, 2000.

8. Ibid.

CHAPTER 19

1. Pontin, Jason, "The Genomic Revolution," *The Red Herring,* May 1998, www.redherring.com.

2. "Joint Statement by President Clinton and Prime Minister Tony Blair of the U.K.," www.pub.whitehouse.gov, 3/14/00. The statement came after a highly publicized breakdown in talks on a collaboration between Celera and HGP.

3. Human Genome Sciences press release, February 2000, www.hgs.com.

4. Fisher, Lawrence, "Surfing the Human Genome," *New York Times,* Sept. 20, 1999.

5. Uetz, P., et al. "A Comprehensive Analysis of Protein-Protein Interactions in *Saccharomyces Cerevigiae,*" *Nature* **403:** 623–627 (2000).

6. www.biospace.com/articles/bio_smart.cfm, Biospace Web site, first posted 4/3/00.

7. Robinson, James, "Entrepreneur Jim Clark gives $150 Million to Bio-X," *Stanford Report,* 10/27/99.

8. Agnew, B., "NIH Plans Bioengineering Initiative," *Science* **200:**1516–1518 (1998).

9. Agrawal, A. "New Institute to Study Systems Biology," *Nature Biotechnology* **17:**743–744 (1999).

10. Lang Jones, J., in *The Experiment Digital Culture,* March 2000.

11. Immunex press release, February 9, 2000, www.immunex.com.

12. Anderson, N.L., Hofmann, J.P., Gemmell, A., and Taylor, J., "Global Approaches to Quantitative Analysis of Gene-Expression Patterns Observed by Use of Two-Dimensional Gel Electrophoresis," *Clin Chem.* **30:**2031–2036 (1984).

13. See Steven Pinker, *The Language Instinct,* HarperPerennial, New York (1995).

14. Ibid.

15. Werner Gitt, *In the Beginning was Information,* translated to English by Dr. Jaap Kies, Christliche Literatur-Verbreitung: Biefeld, Germany (1997).

16. "They deliver their information load to earth, created the chemical bonds for the first molecules of life and heated the molecular soup from which life emerged." Werner Loewenstein, *The Touchstone of Life,* Oxford University Press, New York (1999).

17. Mason, Nick, Gilmour, David, Waters, Roger, and Wright, Rick, *Echoes,* (1971).

APPENDIX

1. The couple who are said to have called the Stanford University Technology Transfer Department regarding access to a patent that broadly covered recombinant DNA methods had no need to be concerned. Their intended procreative activities, a "natural" means of recombining DNA, could not be restricted by a patent.

2. 447 US 303, 1980.

3. US patent 4,736,866.

4. Amgen vs. Chugai Pharmaceuticals, 927 F.2d 1200, 18 USPQ2d 1016 (Fed. Cir.).

5. Frier vs. Reval 984 F2d 1164, 25 USPQ 1601.

6. Fees for filers with less than 500 employees are significantly less than those of larger institutions. For additional savings, self-employed inventors may write and prosecute their own patent. See Pressman, David, *Patent It Yourself,* Nolo Press, Berkeley, California (1997).

7. Ducor, P., "Are Patents and Research Compatible?" *Nature* **387:**13–14 (1997).

8. Kiley, T., "Patents on Random Complementary DNA Fragments," *Science* **257:**915–918 (1992).

9. *Gene Wars*, p.322

10. Report of the NIH Working Group on Research Tools, June 4, 1998, www.nih.gov./news/researchtools/index.htm.

11. See Flores, Mauricio, *Nature Biotechnology* **17:**819–820 (1999).

12. Heller, M.A., and Eisenberg, R.S., "Can Patents Deter Innovation?" *Science* **280:**698–701 (1998).

13. *BMJ* **310:**1351 (1995); *Nature* **375:**168, 268 (1995); *Science* **268:** 487 (1995).

14. See Sagoff, Mark, "Patenting Genes: An Ethical Appraisal," in *Issues in Science and Technology,* p. 37–41, U. Texas at Dallas, Richardson, Texas, (1998).

Index

About the Author

Born in 1964, Gary Zweiger grew up in Connecticut and has lived in Berlin, New York City, Rio de Janeiro, and in (on a sailboat) and around San Francisco Bay. He attended Stanford University, where he majored in biological sciences. He later returned to Stanford for doctoral work in Stanford Medical School's Department of Genetics. His thesis was on DNA methylation and replication in the bacterium *Caulobacter crescentus*. He completed postdoctoral work in Genentech's Department of Molecular Oncology and has also worked in research labs at Schering-Plough's DNAX Molecular Biology Research Institute and Columbia University. He taught biology at Stanford and Columbia University (in Morgan's famous fly room), as well as at the California community schools, Skyline College, and City College of San Francisco.

The author became involved in genomics in 1995 when he was asked to advise investors on opportunities in this emerging field of science and industry. In 1996 he began working at what was then the premier genomics company, Incyte Genomics, and contributed to Incyte's pioneering work in gene discovery, gene expression microarrays, and *in silico* biology. At Incyte the author served as Senior Scientist and Senior Strategic Advisor. In 2000 Gary Zweiger became Director of Business Development and Strategic Planning for Agilent Technologies' new Life Sciences Business Unit. Agilent is a massive spin-off of Hewlett-Packard, a cornerstone at the foundation of northern California's Silicon Valley.

The author is single and lives in Palo Alto, California, with his young daughter, Marissa.